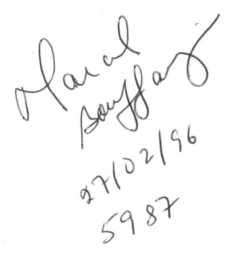

Marcel
Bouffard
27/02/96
5987

Springer Texts in Statistics

Advisors:
Stephen Fienberg   Ingram Olkin

Springer
New York
Berlin
Heidelberg
Barcelona
Budapest
Hong Kong
London
Milan
Paris
Santa Clara
Singapore
Tokyo

Springer Texts in Statistics

Continued at end of book

Ralph O. Mueller

Basic Principles of Structural Equation Modeling

An Introduction to LISREL and EQS

With 25 Illustrations

 Springer

Ralph O. Mueller, Ph.D.
Department of Educational Leadership
Graduate School of Education
    and Human Development
The George Washington University
Washington, DC 20052
USA

On the cover: The artwork is based on an illustration of the basic matrices for structural equation models.

Library of Congress Cataloging-in-Publication Data
Mueller, Ralph O.
    Basic principles of structural equation modeling: an
  introduction to LISREL and EQS/Ralph O. Mueller.
      p.   cm.—(Springer texts in statistics)
    Includes bibliographical references (pp. 216–221) and index.
    ISBN 0-387-94516-4 (hardcover: alk. paper)
    1. LISREL.   2. EQS (Computer file)   3. Path analysis—Data
  processing.   4. Social sciences—Statistical methods.   I. Title.
  II. Series.
  QA278.3.M84   1996
  519.5'35—dc20   95-15043

Printed on acid-free paper.

Production coordinated by Publishing Network and managed by Francine McNeill; manufacturing supervised by Jeffrey Taub.
Typeset by Asco Trade Typesetting Ltd., Hong Kong.
Printed and bound by R.R. Donnelley and Sons, Harrisonburg, VA.
Printed in the United States of America.

9 8 7 6 5 4 3 2 1

ISBN 0-387-94516-4 Springer-Verlag New York Berlin Heidelberg

To Mama and Papa;
to Dan, Paula;
and to Rob:
My family of choice

Preface

During the last two decades, structural equation modeling (SEM) has emerged as a powerful multivariate data analysis tool in social science research settings, especially in the fields of sociology, psychology, and education. Although its roots can be traced back to the first half of this century, when Spearman (1904) developed factor analysis and Wright (1934) introduced path analysis, it was not until the 1970s that the works by Karl Jöreskog and his associates (e.g., Jöreskog, 1977; Jöreskog and Van Thillo, 1973) began to make general SEM techniques accessible to the social and behavioral science research communities. Today, with the development and increasing availability of SEM computer programs, SEM has become a well-established and respected data analysis method, incorporating many of the traditional analysis techniques as special cases. State-of-the-art SEM software packages such as LISREL (Jöreskog and Sörbom, 1993a,b) and EQS (Bentler, 1993; Bentler and Wu, 1993) handle a variety of ordinary least squares regression designs as well as complex structural equation models involving variables with arbitrary distributions.

Unfortunately, many students and researchers hesitate to use SEM methods, perhaps due to the somewhat complex underlying statistical representation and theory. In my opinion, social science students and researchers can benefit greatly from acquiring knowledge and skills in SEM since the methods—applied appropriately—can provide a bridge between the theoretical and empirical aspects of behavioral research. That is, interpretations of SEM analyses can assist in understanding aspects of social and behavioral phenomena if (a) a "good" initial model is conceptualized based on a sound underlying substantive theory; (b) appropriate data are collected to estimate the unknown population parameters; (c) the fit of those data to the a priori hypothesized model is assessed; and (d) if theoretically justified, the initial model is modified appropriately should evidence of lack-of-fit and model misspecification arise. Structural equation modeling thus should be understood as a research *process*, not as a mere statistical technique.

Many social science graduate programs now offer introductions to SEM within their quantitative methods course sequences. Several currently available textbooks (e.g., Blalock, 1964; Bollen, 1989; Duncan, 1975; Hayduk, 1987; Kenny, 1979; Loehlin, 1992) reflect the rapidly increasing interest and advances in SEM methods. However, most treatments of SEM are either outdated and do not provide information on recent statistical advances and modern computer software, or they are of an advanced nature, sometimes making the initial study of SEM techniques unattractive and cumbersome to potential users. The present book was written to bridge this gap by addressing former deficiencies in a readily accessible introductory text.

The foci of the book are basic concepts and applications of SEM within the social and behavioral sciences. As such, students in any of the social and behavioral sciences with a background equivalent to a standard two-quarter or two-semester sequence in quantitative research methods (general ANOVA and linear regression models and basic concepts in applied measurement) should encounter no difficulties in mastering the content of this book. The minimal knowledge of matrix algebra that is required to understand the basic statistical foundations of SEM may be reviewed in this text (Appendix C) or acquired from any elementary linear algebra textbook. The exercises and selected references at the end of each chapter serve to test and strengthen the understanding of presented topics and provide the reader with a starting point into the current literature on SEM and related topics.

Central to the book is the development of SEM techniques by sequentially presenting linear regression and recursive path analysis, confirmatory factor analysis, and more general structural equation models. To illustrate statistical concepts and techniques, I have chosen to use the computer programs LISREL (Jöreskog and Sörbom, 1993a,b) and EQS (Bentler, 1993) to analyze data from selected research in the fields of sociology and counseling psychology. The book can be used as an introduction to the LISREL and EQS programs themselves, although certainly it is not intended to be a substitute for the program manuals (LISREL: Jöreskog & Sörbom, 1993a,b; EQS: Bentler, 1993; Bentler and Wu, 1993). While other computer programs could have been utilized [e.g., Steiger's (1989) EzPATH, Muthén's (1988) LISCOMP, or SAS's PROC CALIS (SAS Institute Inc., 1990)], the choice of introducing LISREL and EQS was based on my perception that they are the most widely used, known, and accepted structural equation programs to date.

The notation used to present statistical concepts is consistent with that in other treatments of SEM (e.g., Bollen, 1989; Hayduk, 1987) and follows the Jöreskog-Keesling-Wiley approach to represent systems of structural equations (e.g., Jöreskog, 1973, 1977; Jöreskog and Sörbom, 1993a; Keesling, 1972; Wiley, 1973). Since all versions of LISREL are based on this approach, the program's matrix-based syntax is easily explained and understood once some underlying statistical theory has been introduced. However, the LISREL package also features an alternative command language, SIMPLIS (Jöreskog and Sörbom, 1993b), which allows the researcher to analyze structural equation models without fully understanding and using matrix representations. While LISREL examples throughout the chapters use the matrix-based syntax to illustrate the statistical foundation and representation of SEM, their equivalents using SIMPLIS are explained in Appendix A.

The EQS language, on the other hand, is built around the Bentler-Weeks model (e.g., Bentler, 1993; Bentler and Weeks, 1979, 1980); this alternative way to represent structural equation systems is not used in this book. Like SIMPLIS, the EQS syntax does not involve matrix specifications and, hence, knowledge of the matrix form of the Bentler-Weeks model is not required to understand and use the EQS program. The package's MS Windows version (Bentler and Wu, 1993) includes the *Build_EQS* option that allows the analyst to create an EQS input file without actually having to write the input program but, instead, by "clicking" on certain options within various dialogue boxes. I feel that knowledge of the program's syntax for model specification is of overall educational value and, thus, discuss actual input files rather than emphasizing the *Build_EQS* option.

Initially, any SEM programming language that does not depend on an understanding of the matrix representation of structural equation models might seem easier than a traditional matrix-based syntax. To acquire an "aesthetic" appreciation of SEM, however, it is necessary to grasp some of the underlying statistical principles and formulations—be it via the approach used here or the one proposed in the Bentler-Weeks model. Getting acquainted with the syntax of the matrix- *and* nonmatrix-based command languages of LISREL and EQS will aid in this endeavor.

In summary, this book was written with four goals in mind: (1) to help users of contemporary social science research methods gain an understanding and appreciation of the main goals and advantages of "doing" SEM; (2) to explicate some of the fundamental statistical theory that underlies SEM; (3) to illustrate the use of the software packages LISREL and EQS in the analysis of basic recursive structural equation models; and, most importantly, (4) to stimulate the reader's curiosity to consult more advanced treatments of SEM.

Organization and Historical Perspective

The book contains three chapters, largely reflecting the historical development of what is known now as the general area of SEM: from a discussion of classical path analysis (Chapter 1) and an introduction to confirmatory factor analysis (Chapter 2) to an exploration of basic analyses of more general structural equation models (Chapter 3). In addition, the book contains five appendices: Appendix A presents all LISREL examples from the chapters in the SIMPLIS command language and serves as an introduction to this alternative LISREL syntax. Appendix B briefly reviews the statistical concepts of distributional location, dispersion, and variable association mainly to explain some of the notation used in this book; some elementary concepts of matrix algebra are presented in Appendix C. Finally, Appendices D and E contain coding information and summary statistics for all variables used in the LISREL and EQS example analyses.

Chapter 1 introduces the language and notation of SEM through a discussion of classical recursive path analysis. Most modern treatments trace the beginnings of SEM to the development of path analysis by Samuel Wright during the 1920s and 1930s (Wright, 1921, 1934). However, it took almost half a century before the method was introduced to the social sciences with articles by Duncan (1966) in sociology, Werts and Linn (1970) in psychology, and Anderson and Evans (1974) in education, among others. Early path analyses were simply ordinary least squares (OLS) regression applications to sets of variables that were structurally related. In many ways, SEM techniques still can be seen as nothing more than the analysis of simultaneous regression equations. In fact, if the statistical assumptions underlying linear regression are met, standard OLS estimation—as available in computer programs such as SPSS, SAS, or BMDP—can be used to estimate the parameters in some structural equation models.

Throughout the review of univariate simple and multiple linear regression in Chapter 1, standard regression notation is replaced by that commonly used in SEM (the Jöreskog-Keesling-Wiley notation system), providing an easy transition into the basic concepts of SEM. The introduction of path diagrams and structural equations leads to the First Law of Path Analysis, which allows for the decomposition of covariances between any two variables in a recursive path model. In turn, such decompositions lead to the definitions of the direct, indirect, and total effects among structurally ordered variables within a specific model. Finally, a discussion and demonstration of the identification problem in recursive path models is included to sensitize the reader to a complex issue that demands attention before SEM analyses can be conducted. Throughout the chapter, theoretical concepts are illustrated by annotated LISREL and EQS analyses of a data set taken from the sociological literature.

In Chapter 2, the inclusion of latent, unobserved variables in structural equation models is introduced from a confirmatory factor analysis (CFA) perspective. Based on Spearman's (1904) discovery of exploratory factor analysis (EFA), several individuals (e.g., Jöreskog, 1967, 1969; Lawley, 1940, 1967; Thurstone, 1947) advanced the statistical theory to represent relationships among observed variables in terms of a smaller number of underlying theoretical constructs, that is, latent variables. It was not until the mid-1950s that a formal distinction was made between EFA and CFA: While the former method is used to search data for an underlying structure, the latter technique is utilized to confirm (strictly speaking, disconfirm) an a priori hypothesized, theory-derived structure with collected data. In this text, only CFA models are discussed; books by Gorsuch (1983), Mulaik (1972), or McDonald (1985) can be consulted for detailed introductions to EFA.

Based on the assumption that most measured variables are imperfect indicators of certain underlying, latent constructs or factors, CFA allows the researcher to cluster observed variables in certain prespecified, theory-driven ways, that is, to specify a priori the latent construct(s) that the observed variables are intended to measure. After estimating the factor loadings (structural coefficients in the regressions of observed on the latent variables) from a variance/covariance matrix of the observed variables, the investigator can and should assess whether or not the collected data "fit" the hypothesized factor structure. Since several issues surrounding the question of how best to assess data-model fit still are unresolved and current answers remain controversial, a detailed discussion of some of the currently utilized data-model fit indices is presented in Chapter 2. The different approaches to data-model fit, as well as other concepts introduced in Chapter 2 (e.g., model modification), are illustrated by (a) an extended analysis of the data set introduced in Chapter 1, and (b) a LISREL and EQS CFA to assess the validity and reliability of a behavior assessment instrument taken from the counseling psychology literature.

Finally, Chapter 3 treats the most general recursive structural equation model presented in this book. The conceptual integration of the classical path analysis model (Chapter 1) and the CFA model (Chapter 2) is used to develop the general structure. During the 1970s, several researchers (e.g., Keesling, 1972; Jöreskog, 1977; Wiley, 1973) succeeded in developing the statistical theory and computational procedures for combining path analysis and CFA into a single statistical framework. The various versions of LISREL, developed by Jöreskog and associates dating back to 1970, continuously incorporated up-to-date statistical advances in SEM. Similarly, EQS (Bentler, 1993) is regarded as one of the most sophisticated and respected software packages for the analysis of a wide variety of structural equation models, including path analytical, CFA, and more general models. In addition, its MS Windows version (Bentler and Wu, 1993) includes many exploratory data analytical tools, such as outlier analysis and other descriptive statistics with various plot options, in addition to inferential techniques, such as dependent and independent t-tests, regression, and ANOVA.

After discussing the specification and identification of general structural equation models that involve latent variables in Chapter 3, the estimation of direct, indirect, and total effect components introduced in Chapter 1 is simplified by introducing a matrix algorithm to estimate the various effect components. Also, two of the most common iterative methods for estimating parameters in general structural equation models are discussed: maximum likelihood (Jöreskog, 1969, 1970; among others) and generalized least squares (Goldberger, 1971; Jöreskog and Goldberger, 1972). While these techniques have important advantages over more traditional methods such as OLS, both methods still depend on the multivariate normality assumption; for data from nonnormal distributions, Browne's (1982, 1984) asymptotically distribution free (ADF) estimators can be obtained from both the LISREL and EQS packages. Following the organization of previous chapters, the concepts presented in Chapter 3 are illustrated with LISREL and EQS analyses of extensions of the models introduced in Chapters 1 and 2 [a demonstration of the ADF method is omitted since it would require either the use of PRELIS (Jöreskog and Sörbom, 1993c) or an analysis of raw data].

A Brief Comment on Causality[1]

It is only natural that the growing attention paid to structural equation modeling—called by some *causal* modeling—has led to a number of criticisms. Baumrind (1983), Cliff (1983), Freedman (1987), and Ling (1983), for example, all presented very insightful perspectives on the dangers associated with utilizing SEM for understanding social science phenomena. Often, however, the reader of such critical views might not recognize immediately which step in the SEM process is being challenged: (a) model conceptualization and specification, (b) identification, (c) parameter estimation, (d) data-model fit assessment, or (e) model modification. Whereas steps (b), (c), and (d) might raise certain statistical questions, steps (a) and (e) are based mostly on some key philosophical assumptions and/or substantive considerations. Duncan (1975) and Wolfle (1985) pointed to the importance and implication of this differentiation: "The study of structural equation models can be divided into two parts: the easy part and the hard part" (Duncan, 1975, p. 149). "The easy part is mathematical. The hard part is constructing causal models that are consistent with sound theory. In short, causal models are no better than the ideas that go into them" (Wolfle, 1985, p. 385). Using this conceptualization, the present book may be seen as an introduction to the "easy" part of SEM. Mastering the "hard" part requires knowledge and understanding of the theoretical, substantive, and philosophical underpinnings of the particular research question(s) as well as an appreciation for the goals and limitations of social science research methodology in general.

One of the most controversial philosophical issues surrounding SEM is the role and interpretation of the concept of causality. A review of major paradigm shifts in the philosophy of science since the British empiricists reveals that "causality" repeatedly has been in and out of favor with philosophers and researchers in the natural and social sciences (see Mulaik, 1987, and references therein). For example, while some contemporary methodologists suggest that "causal explanation is the *ultimate* aim of science" (Shaffer, 1992, p. x; italics added), others insist that "it would be very healthy if more researchers abandon thinking of and using terms such as *cause* and

[1] The comments in this section are taken largely from those expressed in Mueller and Dupuy (1992). The contributions by Paula Dupuy are greatly appreciated and hereby acknowledged.

effect" (Muthén, 1992, p. 82). It seems that this controversy can be explained partially by the arduous challenge facing the applied researcher of translating philosophical concepts into scientific methodologies: The difficult task is using the common-sense notion of causality as a way to explain human experience and finding ways to operationalize the concept for rigorous and meaningful scientific research.

In my view, some of the main criticisms leveled against SEM are closely linked to the particular conceptualizations, definitions, and uses of the concept of causality. While most contemporary operational definitions of causality overlap at least partially, there are differences among them that could have significant impacts on the model conceptualization and interpretation phases of the SEM process. For example, some contributors make the epistemological assumption that nonexperimental research *can* generate causal inferences (e.g., Glymour *et al.*, 1987), whereas the motto "no causation without manipulation" (Holland, 1986, p. 959) implies the inappropriateness of causal modeling in a nonexperimental research setting. Also, researchers disagree on requiring time precedence as a necessary condition for causality: For example, while Kenny (1979) saw cause and effect ordered by time, Marini and Singer (1988) argued that temporal order is not necessary for cause–effect relationships to exist. If time precedence is seen as a necessary condition for causality, only recursive models (models that do not include bidirectional causal relations) are valid. However, if the cause does not need to precede the effect in time, nonrecursive models can be specified (for a classic example, see Duncan *et al.*, 1968). Finally, differences in causality definitions seem to contribute to the discussion over the appropriateness of SEM in exploratory versus confirmatory research modes. Whereas Glymour *et al.* (1987) claimed to have methods that *discover* causal structures, Bollen (1989) sought "only" to *reject* a priori specified models. This difference is basically an epistemological one that can be reconciled only by considering their different sets of causality assumptions.

In conclusion, many experts have explored and sought to clarify the role of causality in scientific research in general and SEM in particular. In this book, I do not attempt to add to these discussions in any meaningful way (except, perhaps, by expressing and explaining at various points throughout the book that SEM should *not* be seen as an exploratory data analysis tool). On the contrary, I admit to somewhat avoiding the issue of causality by using the seemingly more neutral term *structural*—instead of *causal*—when describing relationships among variables that are more than just correlational in nature (in doing so, however, note that it remains difficult to escape the implicit underlying notion of causality). Thus, the following pages deal mainly with an introduction to the "easy" part of SEM, leaving an explanation of the "hard" part to philosophers and other experts. Below, the interested reader is referred to a number of sources that should provide a good entry point into the literature dealing with the philosophical issues surrounding the concept of causality within SEM.

A Note to Instructors

No difficulties should be encountered in covering the content of the book in a one-quarter or one-semester course. When considerably less time is available and/or coverage of the main topics is sought purely from a nontechnical perspective (i.e., when no coverage of basic—but matrix-based—concepts underlying SEM is planned), it is possible to focus mostly on the nonmatrix-based SIMPLIS and EQS illustrations since the examples are relatively self-contained. The more technical parts of the book then serve as a resource for future reference and self-study.

It seems helpful to students to supplement a text such as this with additional readings from the applied research literature that use SEM techniques for data analysis. The LISREL and EQS manuals are good resources for additional examples and, at the very least, should be made available to students for the duration of the course for reference purposes and specific program information. Also, students should be encouraged to consult the more philosophical treatments of SEM (some of which are listed below), in particular those dealing with the role of causality in SEM.

I should mention explicitly that this book does *not* cover certain topics that are, in my opinion, of a somewhat advanced nature and do not need to be part of a *first* exposure to SEM data analysis. Analyses of nonrecursive models, multigroup designs, second-order CFAs, mean structures, and ways of dealing with missing data are not covered here; instead, the reader is referred to more advanced sources such as Bollen (1989). Further, I treat all variables used in the examples as continuous when, in fact, strictly speaking, some are measured on an ordinal scale only [for a detailed discussion on the analysis of ordinal data, consult the manuals accompanying the PRELIS and LISCOMP programs (Jöreskog and Sörbom, 1993c; Muthén, 1988, respectively)].

Of course, I must take responsibility for any errors that are present in the pages to follow. I invite you to contact me if you discover such errors or if you have other comments that could be of value for possible revisions or future editions of the book.

Acknowledgments

Finally, this project depended on the help of many individuals. I thank my former students at The University of Toledo who had to contend with several rough drafts of this book, provided feedback, and kept encouraging me to finish what I had started. Once I felt that the drafts were becoming more presentable, I started to share copies with colleagues whose invaluable input led to the book you see before you. In particular, I thank Peter Bentler and an anonymous reviewer for their in-depth and constructive critiques, and Karl Jöreskog for a personal account of the history of SEM that inspired me greatly. Scientific Software, Inc., and BMDP Statistical Software, Inc., are acknowledged for providing review copies of the LISREL and EQS packages. Of course, Martin Gilchrist, Senior Editor at Springer-Verlag, and Elise Oranges, copyeditor and Director of Publishing Network, deserve much credit for the final product.

Washington, DC   R.O. MUELLER

Recommended Readings

For acquiring a historical perspective on SEM, the following two references might be helpful:

Bentler, P.M. (1980). Multivariate analysis with latent variables: Causal modeling. *Annual Review of Psychology, 31*, 419–456.
Bentler, P.M. (1986). Structural modeling and Psychometrika: An historical perspective on growth and achievements. *Psychometrika, 51*(1), 35–51.

Excellent references for gaining an initial overview and understanding of current controversies surrounding SEM, especially as they relate to the philosophical issue of causality, are the following:

Baumrind, D. (1983). Specious causal attributions in the social sciences. *Journal of Personality and Social Psychology, 45*, 1289–1298.
Bullock, H.E., Harlow, L.L., and Mulaik, S.A. (1994). Causation issues in structural equation modeling research. *Structural Equation Modeling, 1*(3), 253–267.
Mulaik, S.A. (1987). Toward a conception of causality applicable to experimentation and causal modeling. *Child Development, 58*, 18–32.
Shafer, J.P. (Ed.) (1992). *The Role of Models in Nonexperimental Social Science: Two Debates*. Washington, DC: American Educational Research Association.

Contents

Contents

Frequently Used Symbols and Notation

Symbol	Meaning
α (alpha)	Intercept term in a regression or structural equation
β (beta)	Structural effect from an endogenous to another endogenous variable
B (Beta)	Matrix containing the structural effects from endogenous to other endogenous variables
γ (gamma)	Regression or structural coefficient associated with an effect from an exogenous to an endogenous variable
Γ (Gamma)	Matrix containing the structural effects from exogenous to endogenous variables
δ (delta)	Measurement error in observed exogenous variable
ε (epsilon)	Measurement error in observed endogenous variable
ζ (zeta)	Error term associated with an endogenous variable
η (eta)	Latent endogenous variable
Θ_δ (Theta Delta)	Variance/covariance matrix of measurement errors associated with observed exogenous variables
Θ_ε (Theta Epsilon)	Variance/covariance matrix of measurement errors associated with observed endogenous variables

Λ_X **(Lambda X)**	Matrix containing the structural coefficients linking the observed and latent exogenous variables
Λ_Y **(Lambda Y)**	Matrix containing the structural coefficients linking the observed and latent endogenous variables
ξ (ksi)	Latent exogenous variable
ρ (rho)	Pearson's product-moment correlation coefficient
Σ **(Sigma)**	Unrestricted variance/covariance matrix of observed variables
$\Sigma(\theta)$	Model-implied variance/covariance matrix of observed variables
Φ **(Phi)**	Variance/covariance matrix of endogenous variables
Ψ **(Psi)**	Variance/covariance matrix of error terms associated with endogeneous variables

LISREL Abbreviations

Command			Meaning
Main	*Sub*	*Options*	
DA			DAta line
	MA		Matrix to be Analyzed
		CM	Covariance Matrix
	NI		Number of Input variables
	NO		Number of Observations
LA			LAbels for observed variables
LK			Labels for latent exogenous Ksi (ξ) variables
LE			Labels for latent endogenous Eta (η) variables
CM			Covariance Matrix (input)
KM			[K]orrelation Matrix (input)
		FU	FUll
		SY	SYmmetric
ME			MEans
SD			Standard Deviations
SE			SElect line
MO			MOdel line
	NE		Number of latent endogenous Eta (η) variables
	NK		Number of latent exogenous Ksi (ξ) variables

Command			Meaning
Main	Sub	Options	
	NX		Number of observed exogenous X variables
	NY		Number of observed endogenous Y variables
	AL		ALpha
	BE		BEta
	GA		GAmma
	LX		Lambda X
	LY		Lambda Y
	PH		PHi
	PS		PSi
	TE		Theta Epsilon
	TD		Theta Delta
		FI	FIxed
		FR	FRee
		FU	FUll
		DI	DIagonal
		SD	Sub-Diagonal
		SY	SYmmetric
		ZE	ZEro
VA			VAlue
FR			FRee
FI			FIx
OU			OUtput line
		EF	EFfects
		ME	Method of Estimation
		ND	Number of Decimals
		SC	Standardized Completely
		SS	Standardized Solutions

EQS Abbreviations

Abbreviation	Meaning
D	Disturbance
E	Error
F	Factor (latent)
V	Variable (observed)

List of Figures and Tables

Figures

Tables

CHAPTER 1

Linear Regression and Classical Path Analysis

Overview and Key Points

In many ways, structural equation modeling (SEM) techniques may be viewed as "fancy" multivariate regression methods. Some models can be understood simply as a set of simultaneous regression equations. If the statistical assumptions of ordinary least squares (OLS) regression are met, standard OLS estimation as available in general-purpose statistical computer programs such as SPSS, SAS, or BMDP can be used to estimate the structural parameters in these models. This chapter serves to set the stage for presenting general structural equation models in Chapter 3, which include as special cases the regression and path analytical models presented in this chapter and the confirmatory factor analysis models introduced in Chapter 2. For now, the two main purposes are (1) to introduce a general SEM notation system by reviewing some important results in univariate simple and multiple regression, and (2) to discuss the multivariate method of path analysis as a way to estimate direct, indirect, and total structural effects within an a priori specified structural model. Throughout the chapter, examples based on data from a sociological study serve as an introduction to the LISREL and EQS programs. The respective manuals (Jöreskog and Sörbom, 1993a,b; Bentler, 1993; Bentler and Wu, 1993) should be consulted for more detailed programming information. Specifically, eight key points are addressed in this chapter:

1. Univariate linear regression is a method for statistically predicting or explaining a dependent variable by expressing it as a linear and additive function of some relevant independent variable(s).
2. The method of ordinary least squares (OLS), a well-known technique for estimating parameters in a linear regression model from sample data, minimizes the sum of squared differences between observed and predicted

(model-implied) scores of the dependent variable. Correct parameter estimation based on OLS depends on a number of sometimes unrealistic assumptions about the observed variables. Chapter 3 introduces two alternative estimation methods that depend on somewhat less restrictive assumptions.

3. Traditionally, the problem of multicolinearity in its simplest form has been characterized by a high correlation between two or more independent variables in a regression (or structural) equation, which might lead to empirical underidentification of the equation. Erroneous interpretations of the results, mainly due to a lack of stability of coefficients across samples, can follow.

4. Classical path analysis is a multivariate method based on linear regression that allows the researcher to estimate the strengths of the direct structural effect from one variable to another and the indirect effects through intervening variables within the context of an a priori specified path model. Path diagrams are a convenient way to graphically represent the structural relations between variables, and, if certain statistical assumptions are met, OLS regression coefficients may be used to estimate the direct and indirect effects among variables in a particular model.

5. A path analytical model is specified completely by the patterns of 0 and nonzero coefficients in four basic matrices: **Beta** (denoted as **B** or **BE**), a matrix of structural coefficients among endogenous (i.e., dependent) variables; **Gamma** (denoted as Γ or **GA**), a matrix of structural coefficients from exogenous (i.e., independent) to endogenous variables; **Phi** (denoted as Φ or **PH**), a variance/covariance matrix of exogenous variables, and **Psi** (denoted as Ψ or **PS**), a variance/covariance matrix of error terms associated with endogenous variables.

6. The First Law of Path Analysis may be used to decompose the covariances between any two variables in a model into additive components that contain only elements of the basic matrices **B**, Γ, Φ, and Ψ.

7. The direct effect (*DE*) of an independent on a dependent variable in a path model is defined as the associated structural coefficient that links the two variables. A particular indirect effect of one variable on another through one or more intervening variables is expressed by the product of the coefficients along a specific structural chain of paths that link the two variables. The sum of all possible particular indirect effects for a pair of variables is the total indirect effect (*IE*), and the sum of the direct and total indirect effects is defined as the total structural effect (*TE*).

8. Before data analysis, the identification status of a model (just, over, or under-identified) must be considered. In an underidentified model, for example, the estimation of some parameters in the model is impossible due to insufficient variance/covariance information from the observed variables.

The data used for the LISREL and EQS examples in this chapter are taken from a sociological study by Mueller (1988). Summary statistics from this investigation are used here to assess the influence of parents' on respondent's socioeconomic status (SES), although the main question of the original study was how strong of an impact the prestige of the college one attends has on one's income 5 years after graduation. Mueller (1988) used linear regression and path analytical methods to analyze data from 3094 male and 3833 female individuals obtained from the 1971 ACE/UCLA Freshman Survey with the 1979–1980 HERI Follow-Up Survey and the accompanying HEGIS data. Appendix D contains summary statistics (means, standard deviations, and product–moment correlations) for all variables in the analysis and a list of how each variable is coded. For the examples in this chapter, only data from the male subsample ($n = 3094$) on the following five variables are considered: father's educational level (*FaEd*), respondent's high school rank (*HSRank*) and degree aspiration (*DegreAsp*), college selectivity (*Selctvty*), and respondent's highest held academic degree (*Degree*).

Linear Ordinary Least Squares Regression

Simple and multiple linear regression techniques have been used extensively in the social science literature as data analytical tools for the purposes of prediction and variance explanation. Usually, when referring to regression analyses, the statistical method of how unknown population parameters are estimated from sample data is assumed to be ordinary least squares (OLS). Later in this book, alternative estimation techniques are introduced that do not depend on some of the stringent assumptions underlying the OLS method. For example, prediction equations soon can include unobserved or latent factors that are measured by imperfect indicator variables; that is, measurement error in observed variables can be taken into account when assessing the influence of a set of independent variables on one or more dependent variables. Also, methods are introduced in this chapter that allow for assessing *indirect* effects of independent on dependent variables via intervening factors, as opposed to the estimation of just the *direct* effects within a classical regression framework. For now, however, we concentrate on reviewing some selected topics and issues in simple and multiple OLS regression with the main purposes of (a) introducing some notation that is needed for the more general structural equation modeling topics of later sections and chapters, and (b) familiarizing the reader with some basic LISREL and EQS syntax. Exercises and suggestions on where to find in-depth treatments of regression methods are listed at the end of the chapter.

Simple Linear Regression

The prediction of an individual's socioeconomic status (SES) based on parents' SES might be a problem of possible interest to sociologists. Individual's highest held academic degree (*Degree*) may be used as an indicator variable of respondent's SES and here represents the dependent variable. Father's educational level (*FaEd*), a possible indicator variable of parents' SES, serves as the independent variable. When a researcher is interested in predicting a dependent variable like highest held academic degree ($Y = Degree$) from a single independent variable like father's educational level ($X = FaEd$), the most often used univariate (one dependent variable) statistical technique is *simple* (one independent variable) linear regression. The assumed linear relationship between Y and X can be expressed by the following statistical model, called the *regression equation*:

$$Y_k = \alpha + \gamma_{YX} X_k + \zeta_k, \tag{1.1}$$

where the subscript k denotes the kth individual, $k = 1, \ldots, N$; Y is the dependent variable, e.g., *Degree*; X indicates the independent variable, e.g., *FaEd*; α (*alpha*) denotes the intercept term; γ_{YX} (*gamma*), is the regression coefficient; and ζ_k (*zeta*) indicates random error associated with the kth individual's Y-score.

The estimation of the population intercept term α and regression coefficient γ_{YX} in equation (1.1) from sample data on n individuals often is accomplished by the method of *ordinary least squares* (OLS) although, as discussed in Chapter 3, there are other, more general estimation techniques available. When the regression equation in (1.1) is estimated from data on n individuals, it takes the form

$$\hat{Y}_k = \hat{\alpha} + \hat{\gamma}_{YX} X_k, \tag{1.2}$$

where $\hat{Y}_k$ denotes the predicted, or model-implied, value of Y_k, and $\hat{\alpha}$ and $\hat{\gamma}_{YX}$ are sample estimates of α and γ_{YX}, respectively. The error in predicting Y-scores from observed X-scores, $\hat{\zeta}_k = Y_k - \hat{Y}_k$, is a sample estimate of ζ_k in equation (1.1) and is used in obtaining the OLS sample estimates.

Specifically, the method of OLS minimizes the sum of squared differences between the observed and model-implied Y-values,

$$\sum_{k=1}^{n} \hat{\zeta}_k^2 = \sum_{k=1}^{n} (Y_k - \hat{Y}_k)^2. \tag{1.3}$$

Elementary differential calculus shows that the OLS estimates for γ_{YX} and α in equation (1.1) can be calculated by

$$\hat{\gamma}_{YX} = \hat{\sigma}_{YX} / \hat{\sigma}_X^2 \tag{1.4}$$

and

$$\hat{\alpha} = \hat{\mu}_Y - \hat{\gamma}_{YX} \hat{\mu}_X, \tag{1.5}$$

where $\hat{\sigma}_{YX}$ is the sample estimate of the covariance between variables Y and X; $\hat{\sigma}_X^2$ denotes the estimate of the variance in variable X; and $\hat{\mu}_Y$ and $\hat{\mu}_X$ are sample estimates of the means of variables Y and X, respectively.

Using equations (1.4) and (1.5) directly, or any of the standard OLS regression programs (e.g., SPSS, SAS, or BMDP) with the data from the male subsample given in Appendix D, $\hat{\gamma}_{YX} = 0.082$ and $\hat{\alpha} = 4.227$. Thus, the estimated regression equation (1.2) becomes

$$\hat{Y} = 4.227 + 0.082X,$$

indicating that for a 1-point increase on the *FaEd* scale, one can expect, on average, an increase of 0.082 on the *Degree* scale. It might be tempting to interpret the estimate of the intercept, $\hat{\alpha} = 4.227$, as the predicted value on the *Degree* scale that corresponds to a value of 0 on the *FaEd* scale. The coding schema in Appendix D, however, indicates that 0 is not in the range of possible values for the variable *FaEd*, and hence, a substantive interpretation of the intercept term is not appropriate.

If both variables Y and X are standardized, equations (1.4) and (1.5) imply that $\hat{\gamma}_{YX} = \hat{\rho}_{YX}$ and $\hat{\alpha} = 0$, so that equation (1.2) reduces to

$$\hat{Y} = \hat{\rho}_{YX}X,$$

where $\hat{\rho}_{YX}$ denotes the sample estimate of the Pearson product–moment correlation coefficient between variables Y and X. In general, standardized regression coefficients can be computed from the unstandardized (metric) coefficients by

$$\hat{\gamma}_{(stand)} = \hat{\gamma}_{(metric)}(\hat{\sigma}_X/\hat{\sigma}_Y). \tag{1.6}$$

Thus, it can be shown that $\hat{\gamma}_{(stand)} = 0.129$ for the above example, indicating that for a one standard deviation increase on the *FaEd* scale, an average 0.129 standard deviation increase on the *Degree* scale is predicted.

Finally, consider the correlation between observed Y-scores and predicted (model-implied) Y-scores when the variables are standardized. Using the rules of covariance algebra (see Appendix B), one observes that

$$\hat{\rho}_{Y\hat{Y}} = \hat{\rho}_{Y(\hat{\rho}_{YX}X)} = \hat{\rho}_{YX}^2. \tag{1.7}$$

Commonly, the right side of equation (1.7) is referred to as the *coefficient of determination*. This measure represents the proportion of variance in the Y-variable that can be attributed to the variability in the independent variable X, and vice versa. For the above example, $\hat{\rho}_{YX}^2 = 0.017$, indicating that, for male respondents, only about 1.7% of the variability in *Degree* is accounted for by the variability in *FaEd*. Alternatively, this low coefficient of determination can be interpreted as an indication that the hypothesized relationship between *Degree* and *FaEd* in equation (1.1) is not supported strongly by the observed data.

The null hypothesis that a coefficient of determination is equal to 0 can be statistically tested by the F-ratio

$$F = \frac{\hat{\rho}_{YX}^2/NX}{(1 - \hat{\rho}_{YX}^2)/(n - NX - 1)} \tag{1.8}$$

with $df_{num} = NX$ and $df_{den} = (n - NX - 1)$, where NX indicates the number of independent variables in the regression equation. For the simple regression example, $NX = 1$ and equation (1.8) becomes

$$F = \frac{0.017/1}{(1 - 0.017)/(3094 - 1 - 1)} = 53.473$$

with $df_{num} = 1$ and $df_{den} = 3092$. Thus, even though the proportion of explained variability in the dependent variable (*Degree*) is small, the high statistical power of the test (due to the large sample size) leads to the conclusion that the coefficient of determination is significantly different from 0. That is, father's educational level explains a statistically significant (albeit small) portion of the variability in respondent's highest held academic degree.

LISREL EXAMPLE 1.1: SIMPLE LINEAR REGRESSION OF ACADEMIC DEGREE ON FATHER'S EDUCATIONAL LEVEL. The next two sections present the regression example of respondent's highest held degree (*Degree*) on father's educational level (*FaEd*) using the structural equation modeling program LISREL (Jöreskog and Sörbom, 1993a). Statistical software packages such as SPSS, SAS, or BMDP might be more suitable for univariate OLS regression analyses since they include certain diagnostic features that generally are not available in SEM packages. The analysis is included only to introduce some elementary LISREL syntax.

Beginning with Version 8 of LISREL, two very different command languages are available, termed LISREL (LInear Structural RELations) and SIMPLIS (SIMPle LISREL) by the program authors. While the LISREL command language is based on a matrix representation of general structural equation models, SIMPLIS language does not require the specification of model matrices. *Within this book, all LISREL examples use the former command language, and Appendix A presents corresponding analyses using the latter language.* Below, the input file for the simple linear regression example is explained in detail, and selected output verifies that LISREL's parameter estimates of the regression coefficient γ_{YX} and the intercept term α are equal to corresponding estimates computed from equations (1.4) and (1.5).

LISREL Input. In general, LISREL commands are certain line names followed by keywords and/or options. Every LISREL program—from the relatively simple one presented in Table 1.1 to the more complex programs discussed in Chapter 3—has a similar structure. Each consists of four main sections: (1) program title, (2) data input, (3) model specification, and (4) analysis output. The program title section allows the user to specify an identifying title of the program. The data input section contains information such as the number of input variables and observations, explanatory variable labels, and the form and source of the input data (e.g., raw data or a correlation matrix with means and standard deviations, either given as part of the program or placed in an external data file). In the model specification section,

Table 1.1. LISREL Input File for a Simple
Linear Regression Example

1	Example 1.1. Simple Linear Regresssion
2	DA NI = 2 NO = 3094
3	LA
4	Degree FaEd
5	KM SY
6	1
7	.129 1
8	ME
9	4.535 3.747
10	SD
11	.962 1.511
12	MO NY = 1 NX = 1 AL = FR
13	OU SC ND = 3

the user must provide information regarding the particular form of the statistical model that is being analyzed. Finally, the analysis output section allows the user to customize LISREL's output by requesting, for example, the standardized solutions for the estimated model parameters or that the results are rounded to more than two decimals, the default in LISREL.

The input file for the regression of *Degree* on *FaEd* is shown in Table 1.1. The program lines are numbered for reference purposes only; that is, the numbers in the left margin of Table 1.1 are not part of the actual input program. We now discuss each program line in detail.

Line 1: Title line (optional). Any title can be specified as long as it does not start with the upper- or lowercase character combination "da".

Line 2: DAta line (required). For the simple regression example, the Number of Input variables (NI) and the Number of Observations (NO) need to be specified. Two variables are considered in the present example (*Degree* and *FaEd*), and the analysis is based on $n = 3094$ male respondents; thus NI = 2 and NO = 3094.

Lines 3 and 4: LAbel lines (optional). Descriptive names, separated by blank spaces, can be given to the input variables. The labels should not be longer than eight characters. The order of labels must correspond to the order of variables in the input matrix, which is specified on lines 6 and 7.

Lines 5 through 7: Data input lines (required). LISREL can read raw data or summary statistics like a [K]orrelation Matrix (KM) or a Covariance Matrix (CM). A matrix can be entered either in FUll (FU) or SYmmetric (SY) format. With the latter option, only values on and below the main diagonal of the input matrix need to be specified. For the current example, a correlation matrix in symmetric format (KM SY) is used as input data.

Lines 8 through 11: MEans and Standard Deviations lines (conditional). For the example, inclusion of sample MEans (ME) of the two input variables in the LISREL input file is required since the calculation of the intercept term $\hat{\alpha}$ involves the means of *Degree* and *FaEd* [see equation (1.5)]. If an interpretation of the intercept term is not of interest to the researcher or is not appropriate, the ME lines do not need to be included. The estimation of α is included here for demonstration purposes only since, in the context of the example, a substantive interpretation of the intercept term is not appropriate.

The Standard Deviations (SD) of the variables *Degree* and *FaEd* are specified next since parameter estimates should be calculated from a variance/covariance matrix [see equation (1.4)] rather than a correlation matrix. LISREL automatically computes the covariance matrix from the input correlation matrix on lines 6 and 7 and the standard deviations on Line 11. If raw data or a covariance matrix is used as input, the SD line is not required.

Line 12: MOdel line (required). The specific structure of the statistical model is given on the MO line. For the present example, only two types of entries are necessary: (1) the number of dependent and independent variables, Y and X, respectively; and (2) an indication of whether or not the intercept term should be estimated. Presently, *Degree* is the only dependent variable ($NY = 1$) and *FaEd* is the single independent variable ($NX = 1$). The statement AL = FR indicates that the intercept parameter α (ALpha), is FRee— that is, to be estimated—rather than FIxed to 0, the default in LISREL.

Line 13: OUtput line (required). On this line the researcher can specify several options to customize the LISREL output. For example, the LISREL keyword SC (Standardized Completely) requests standardized coefficients and the statement ND = 3 sets the Number of Decimals of the results equal to 3, rather than to the default value of 2.

Selected LISREL Output. Table 1.2 contains selected output from the analysis produced by the LISREL input program in Table 1.1. First, the variance/covariance matrix and means of the variables *Degree* and *FaEd* that are used to compute the estimates of the regression coefficient γ_{YX} and the intercept α are presented. Next, the LISREL estimate of the unknown regression coefficient ($\hat{\gamma}_{YX} = 0.082$) is presented. The associated standard error ($SE = 0.011$) and t-value ($t = 7.234$)—the estimate divided by its standard error—are printed directly below the estimate. If a reported t-value is greater than a certain critical value, the null hypothesis that the associated parameter is equal to 0 is rejected. As a rule of thumb for large samples, t-values greater than 2 generally are taken to indicate statistical significance (for a two-tailed test at the .05 level of significance, the critical value associated with a t-distribution with $df = 60$ is 2.00 and decreases in value as the degrees of freedom increase). Thus, in the example, the regression coefficient ($\hat{\gamma}_{YX} = 0.082, t = 7.234$) is significantly different from 0. Note that LISREL uses maximum likelihood (ML) as the default method to estimate the parameters. This technique will be discussed in Chapter 3. For regression models meeting OLS assumptions, parameter estimates based on ML are equivalent to those based on OLS estimation. This can be verified by comparing the

Table 1.2. Selected LISREL Output of a Simple Linear Regression
Example

COVARIANCE MATRIX TO BE ANALYZED

	Degree	FaEd
Degree	0.925	
FaEd	0.188	2.283

MEANS

Degree	FaEd
4.535	3.747

LISREL ESTIMATES (MAXIMUM LIKELIHOOD)

GAMMA

	FaEd
Degree	0.082
	(0.011)
	7.234

SQUARED MULTIPLE CORRELATIONS FOR STRUCTURAL EQUATIONS

Degree
0.017

ALPHA

Degree
4.227
(0.046)
92.153

STANDARDIZED SOLUTION

GAMMA

	FaEd
Degree	0.129

LISREL estimates in Table 1.2 to those presented in the previous discussion
on OLS regression.

The coefficient of determination ($\hat{\rho}^2_{YX} = 0.017$) is presented next, followed
by a statistically significant estimate of the intercept term ($\hat{\alpha} = 4.227$,
$SE = 0.046, t = 92.153$). Finally, the standardized estimate of the regression
coefficient is given ($\hat{\gamma}_{(stand)} = 0.129$), which is equal to the correlation between
the two variables *Degree* and *FaEd* [see equation (1.4)]. Substantive interpretations of the results in Table 1.2 are the same as those reached during the
above discussion of the OLS estimates.

EQS EXAMPLE 1.1: SIMPLE LINEAR REGRESSION OF ACADEMIC DEGREE ON
FATHER'S EDUCATIONAL LEVEL. The simple regression of the dependent vari-

able *Degree* on the independent variable *FaEd* also can be analyzed with EQS, Version 4 (Bentler, 1993). EQS is a user-friendly and comprehensive SEM software package based on the Bentler-Weeks model to represent systems of structural equations (e.g., Bentler and Weeks, 1979, 1980; or see the EQS manual, Chapter 10). In this book, the Jöreskog-Keesling-Wiley (e.g., Jöreskog 1973, 1977; Keesling, 1972; Wiley, 1973; or see the LISREL manual, Chapter 1) notational system is used, which might lead to initial difficulties in becoming comfortable with the EQS syntax. However, since the EQS command language does not require matrix representations of structural equation models, the differences between the EQS and LISREL languages should pose little or no confusion for the user.

EQS Input. An EQS input program for the regression of *Degree* on *FaEd* is presented in Table 1.3. Line numbers at the left of the table again are included for reference purposes only and are not part of the actual input file. In general, EQS requires model specification in several main sections, each beginning with a slash (/) followed by a keyword naming the particular section. All information within sections—except in the data input portion of the program—must be separated by a semicolon (;). More specifically, EQS programs contain portions pertaining to (a) a suitable program title; (b) model specification [e.g., number of observations and variables, type of input matrix and matrix to be analyzed, and type of estimation method (the default is maximum likelihood as in LISREL)]; (c) the particular model equations and an indication of the parameters that are to be estimated by the program (e.g., regression coefficients and certain variances or covariances); (d) input

Table 1.3. EQS Input File for a Simple Linear
Regression Example

1	/TITLE
2	Example 1.1. Simple Linear Regression
3	/SPECIFICATIONS
4	Cases = 3094; Variables = 2; Matrix = Correlation;
5	/LABELS
6	V1 = FaEd; V2 = Degree;
7	/EQUATIONS
8	V2 = *V1 + E2;
9	/VARIANCES
10	V1 = *; E2 = *;
11	/MATRIX
12	1
13	.129 1
14	/STANDARD DEVIATIONS
15	1.511 .962
16	/END

data (e.g., correlation or variance/covariance matrix, means, standard deviations); (e) specification of particular information to be included in the output; and (f) an indication of the end of the input program. Consult the program's manual (Bentler, 1993) for detailed syntax information.

Lines 1 and 2: /TITLE section (optional). Following the keyword /TITLE, any program title can be specified on the next line.

Lines 3 and 4: /SPECIFICATIONS section (required). The current example involves $n = 3094$ individuals (Cases = 3094;) and two variables (Variables = 2;). Furthermore, a correlation matrix is used for data input (Matrix = Correlation;). Each of these specifications ends with a semicolon (;) as is required by EQS.

Lines 5 and 6: /LABELS section (optional). Following the keyword /LABELS, descriptive names can be given to the input variables. According to the Bentler-Weeks model, in EQS observed Variables (both independent and dependent) are denoted by the letter V and are numbered according to their order of input. That is, the statements "V1 = FaEd;" and "V2 = Degree;" (both end with a semicolon) provide the labels *FaEd* and *Degree* to the two input variables V1 and V2.

Lines 7 and 8: /EQUATIONS section (required). The keyword /EQUATIONS indicates that, following this line, the equation(s) of the model is (are) specified. Under the Bentler-Weeks model, Error terms associated with dependent variables [ζ in equation (1.1)] are denoted by the letter E. Here, V1 (*FaEd*) is the independent variable and V2 (*Degree*) is the dependent variable with E2 denoting the error term associated with V2. Finally, parameters that are being estimated by the program are indicated by asterisks (*). Thus, Line 8 of the input program in Table 1.3 specifies that the regression coefficient associated with the independent variable V1 is to be estimated by EQS.

Due to the interpretation problems noted in previous sections, the intercept term α in equation (1.1) is not specified to be estimated in the EQS input file. Later in this chapter the assumption is made that variables are measured as deviation scores from their means, so that the estimation of intercept terms becomes unnecessary. Thus, estimation of these constants with EQS is omitted; see Bentler (1993, Chapter 8) for a discussion on EQS analyses of models that require the estimation of intercept terms.

Lines 9 and 10: /VARIANCES section (required). All hypothesized non-zero variances and covariances of independent variables need to be specified as parameters to be estimated (the reason for this will become apparent during the later discussion on path analysis). In the Bentler-Weeks model, the error terms associated with dependent variables are considered independent variables with an implied constant regression coefficient set to 1.0; see equation (1.1). Therefore, the variances of V1 (*FaEd*) and E2 [error term associated with V2 (Degree)] need to be estimated as indicated by asterisks (*) on line 10 of the input file. Again, statements within this section end with semicolons.

Lines 11 through 13: /MATRIX section (conditional). Unless raw data in a separate data file are used as input for an analysis, specification of an input matrix of summary statistics is required in EQS. Line 4 indicates that a correlation matrix serves as input (Matrix = Correlation;). Following the /MATRIX command, program lines 12 and 13 specify a lower triangular matrix containing the correlation(s) among the observed variables. Note that the information in this section is *not* separated by semicolons.

Lines 14 and 15: /STANDARD DEVIATIONS section (conditional). Since a correlation matrix serves as input but parameter estimation in general should be based on an analysis of a variance/covariance matrix [see equation (1.4)], EQS requires the standard deviations of the input variables. This information, together with the correlation matrix specified in the previous section, is used to compute the appropriate variance/covariance matrix. Following the /STANDARD DEVIATIONS keyword, standard deviations of the variables V1 (*FaEd*) and V2 (*Degree*) are given. Again, information in this data input section is not separated by semicolons.

Line 16: /END program section (required). Every EQS program file must end with this keyword.

Selected EQS Output. Table 1.4 contains selected parts of the EQS output file. The variance/covariance matrix among the variables *FaEd* and *Degree*

Table 1.4. Selected EQS Output of a Simple Linear Regression Example

---

COVARIANCE MATRIX TO BE ANALYZED: 2 VARIABLES
(SELECTED FROM 2 VARIABLES), BASED ON 3094 CASES.

		FaEd V1	Degree V2
FaEd	V1	2.283	
Degree	V2	0.188	0.925

MAXIMUM LIKELIHOOD SOLUTION (NORMAL DISTRIBUTION
THEORY) MEASUREMENT EQUATIONS WITH STANDARD ERRORS
AND TEST STATISTICS

Degree = V2 =   .082*V1 + 1.000 E2
                  .011
                  7.235

VARIANCES OF INDEPENDENT VARIABLES

V		F		E		D
V1 − FaEd	2.283*	I		I E2 − Degree	.910* I
	.058	I	I		.023 I
	39.326	I	I		39.326I
		I	I		I

STANDARDIZED SOLUTION:

Degree = V2 =   .129*V1 + .992 E2

---

calculated by EQS is printed followed by the estimated regression equation. Directly beneath the estimated regression coefficient ($\hat{\gamma} = 0.082$), EQS prints the associated standard error (0.011) and t-value (7.235). A listing of the variances of the independent variables V1 (*FaEd*) and E2 (error term associated with *Degree*) are presented in the output section entitled VARIANCES OF INDEPENDENT VARIABLES (the two empty columns labeled as F and D are explained in later chapters). Their respective standard errors and t-values again are printed directly under the estimates.

The end of the output file includes the standardized regression equation with $\hat{\gamma}_{(stand)} = 0.129$. Note that EQS does not explicitly print a coefficient of determination $\hat{\rho}^2$ for the regression equation of *Degree* on *FaEd*. However, since the standardized coefficient (0.992) associated with the error term E2 is given by $\sqrt{(1 - \hat{\rho}^2)}$, it can be used to compute $\hat{\rho}^2$: Subtracting the square of the standardized coefficient associated with the error term ($0.992^2 = 0.984$) from 1 ($1 - 0.984 = 0.016$) gives the desired estimate $\hat{\rho}^2$. Finally, all results based on the EQS analysis are equal (within rounding error) to corresponding LISREL results in Table 1.2.

Multiple Linear Regression

Following a similar reasoning underlying the simple regression model, it might be desirable to partially predict or explain a dependent variable Y from two or more independent variables $X_j, j = 1, \ldots, NX$. Extending the research question introduced in the previous section, one might hypothesize that an individual's socioeconomic status (SES) not only depends on the parents' SES but also on motivational level and prestige of the college attended. As before, the variables highest held academic degree ($Y = Degree$) and father's educational level ($X_1 = FaEd$) could be used as indicator variables of respondent's SES and parents' SES, respectively. In addition, degree aspirations ($X_2 = DegreeAsp$) and college selectivity ($X_3 = Selctvty$; measured by average SAT scores) might be appropriate indicator variables of respondent's motivation and college prestige. With these assumptions, the revised regression equation of highest held academic degree (Y) becomes

$$Y_k = \alpha + \gamma_{Y1} X_{1k} + \gamma_{Y2} X_{2k} + \gamma_{Y3} X_{3k} + \zeta_k, \qquad (1.9)$$

where α represents the Y-intercept term; γ_{Yj} denotes the regression coefficient associated with variable $X_j, j = 1, \ldots, NX$; ζ_k is the random error associated with the kth individual, $k = 1, \ldots, N$; and NX denotes the number of independent variables X_j (here, $NX = 3$).

The general statistical model for such a *multiple* (more than one independent variable) linear regression can be written as

$$Y_k = \alpha + \gamma_{Y1} X_{1k} + \gamma_{Y2} X_{2k} + \cdots + \gamma_{YNX} X_{NXk} + \zeta_k, \qquad (1.10)$$

where Y denotes the dependent variable, X_1 through X_{NX} indicate the independent variables, α is the intercept term, γ_{Y1} through γ_{YNX} are the regression coefficients, and ζ_k is the random error associated with Y_k.

Alternatively, a matrix formulation of equation (1.10) provides a convenient and compact way to represent general univariate multiple linear regression models. Equation (1.10) can be re-written as

$$\mathbf{Y} = \boldsymbol{\alpha} + \mathbf{X}\boldsymbol{\Gamma} + \boldsymbol{\zeta}, \tag{1.11}$$

where $\mathbf{Y}$ is a $(N \times 1)$ column vector of Y-scores, $\boldsymbol{\alpha}$ is a $(N \times 1)$ column vector of intercept terms (constant across the N individuals), $\mathbf{X}$ is a $(N \times NX)$ data matrix of X-scores, $\boldsymbol{\Gamma}$ **(Gamma)** denotes a $(NX \times 1)$ column vector of regression coefficients, and $\boldsymbol{\zeta}$ **(Zeta)** is a $(N \times 1)$ column vector of error terms. That is,

$$\begin{bmatrix} Y_1 \\ Y_2 \\ \vdots \\ Y_N \end{bmatrix} = \begin{bmatrix} \alpha \\ \alpha \\ \vdots \\ \alpha \end{bmatrix} + \begin{bmatrix} X_{11} & X_{21} & \cdots & X_{NX1} \\ X_{12} & X_{22} & \cdots & X_{NX2} \\ \vdots & \vdots & \ddots & \vdots \\ X_{1N} & X_{2N} & \cdots & X_{NXN} \end{bmatrix} \begin{bmatrix} \gamma_{Y1} \\ \gamma_{Y2} \\ \vdots \\ \gamma_{YNX} \end{bmatrix} + \begin{bmatrix} \zeta_1 \\ \zeta_2 \\ \vdots \\ \zeta_N \end{bmatrix}. \tag{1.12}$$

The statistical estimation of elements in $\boldsymbol{\Gamma}$ in equation (1.11) based on data from n individuals can be accomplished by applying the OLS (ordinary least squares) method. In brief, the OLS criterion to compute $\hat{\boldsymbol{\Gamma}}$, the sample estimate of $\boldsymbol{\Gamma}$, is to minimize

$$\mathbf{Z}'\mathbf{Z} = \sum_{k=1}^{n} \hat{\zeta}_k^2, \tag{1.13}$$

where the matrix $\mathbf{Z}$ [sample estimate of $\boldsymbol{\zeta}$ in equation (1.11)] consists of elements $\hat{\zeta}_k = Y_k - \hat{Y}_k$ [sample estimates of ζ_k in equation (1.10)]. Using this criterion, $\hat{\boldsymbol{\Gamma}}$ is given by

$$\hat{\boldsymbol{\Gamma}} = (\mathbf{X}'\mathbf{X})^{-1}\mathbf{X}'\mathbf{Y}, \tag{1.14}$$

and the estimate of the intercept term α can be computed as

$$\hat{\alpha} = \hat{\mu}_Y - \hat{\gamma}_{Y1}\hat{\mu}_{X_1} - \hat{\gamma}_{Y2}\hat{\mu}_{X_2} - \cdots - \hat{\gamma}_{YNX}\hat{\mu}_{X_{NX}}. \tag{1.15}$$

Inferences based on OLS estimation depend on a number of statistical assumptions:

1. The independent variables X_j are measured with no or negligible error.
2. The relationship between the dependent variable Y and the independent variables X_j is linear.
3. The error terms in $\boldsymbol{\zeta}$ in equation (1.11) (a) have a mean of 0 and a constant variance across observations; (b) are independent, i.e., uncorrelated across observations; and (c) are uncorrelated with the independent variables.

When these assumptions are violated, regression coefficients based on equation (1.14) might be under- or overestimated, and increased probabilities of Type I errors might be present during hypothesis testing (e.g., see Myers, 1986).

As an example of multiple OLS regression, consider again the task of predicting highest held academic degree ($Y = Degree$) from father's educational level ($X_1 = FaEd$), respondent's degree aspirations ($X_2 = DegreAsp$), and college selectivity ($X_3 = Selctvty$). Using a standard regression program and descriptive data from the male subsample ($n = 3094$) in Appendix D, the regression equation of $Degree$ on $FaEd$, $DegreAsp$, and $Selctvty$ for males [equation (1.9)] can be estimated by

$$\hat{Y} = \hat{\alpha} + \hat{\gamma}_{Y1}X_1 + \hat{\gamma}_{Y2}X_2 + \hat{\gamma}_{Y3}X_3 = 3.170 + 0.029X_1 + 0.195X_2 + 0.095X_3.$$

Again note that the estimated intercept term $\hat{\alpha}$ has a substantive meaning only when the value of 0 is included in the range of all X_j-values (which is not the case in the present example; see Appendix D for the variable codes). A metric (unstandardized) regression coefficient $\hat{\gamma}_{Yj}$ is interpreted as the estimated amount of unit change in the dependent variable Y when X_j is increased by one unit and all other independent variables are held constant. For example, the coefficient $\hat{\gamma}_{Y2} = 0.195$ for the variable $DegreAsp$ in the above regression equation can be interpreted as the estimated increase on the $Degree$ scale when $DegreAsp$ is increased by one unit and the variables $FaEd$ and $Selctvty$ are held constant.

Standardized regression coefficients can be calculated as before [see equation (1.6)]; that is,

$$\hat{\gamma}_{(stand)Yj} = \hat{\gamma}_{(metric)Yj}(\hat{\sigma}_{X_j}/\hat{\sigma}_Y). \tag{1.16}$$

The Y-intercept $\hat{\alpha}$ becomes 0 when standardized scores are used for the analysis [equation (1.15)]. The standardized coefficient associated with variable X_j in equation (1.16) is the estimated amount of standard deviation change in the dependent variable Y when X_j is increased by one standard deviation and all other independent variables are held constant. From the data in Appendix D and using equation (1.16), it can be verified that standardized weights in the example are 0.045, 0.206, and 0.196 for the independent variables $FaEd$, $DegreAsp$, and $Selctvty$, respectively. These coefficients might be helpful when assessing the relative "importance" of the independent variables in explaining the dependent variable: It might be concluded that respondent's degree aspiration has the most influence ($\hat{\gamma}_{(stand)Y2} = 0.206$) and father's educational level the least influence ($\hat{\gamma}_{(stand)Y1} = 0.045$) on highest held academic degree. In general, however, the interpretation of standardized regression weights for the purpose of assessing relative variable importance can be problematic since these coefficients can be very unstable from sample to sample (e.g., Huberty and Wisenbaker, 1992).

Similar to the simple regression result given in equation (1.7), in multiple regression the correlation between observed and predicted Y-scores, $\hat{\rho}_{Y\hat{Y}}$, is equal to the square of the multiple correlation coefficient R between Y and the linear combination of the NX independent variables, X_j. This coefficient of determination, R^2, represents the amount of variance in the dependent variable explained by the set of independent variables and is tested by the same F-ratio as in the simple case [equation (1.8)]. Alternatively, it can be thought of as a measure of "data-model fit" in the sense that a small value of R^2 indicates that the statistical model [the regression equation in equations (1.10) and (1.11)] cannot be confirmed based on the observed data. On the other hand, large coefficients of determination may be used as evidence that the data are consistent with the hypothesized linear relationship between the dependent and independent variables.

It also must be noted that R^2 is a *biased* estimator, that is, its expected value is not equal to the corresponding population parameter. Thus, R^2-values should be interpreted only descriptively, within the context of a particular data set. If an unbiased estimate of the population coefficient of determination is needed, an adjusted R^2, R_{adj}^2, can be computed by

$$R_{adj}^2 = 1 - \frac{(n-1)}{(n - NX - 1)}(1 - R^2).$$

This "shrinkage" formula (Cohen and Cohen, 1983, pp. 105–107) adjusts the value of the coefficient of determination downward (only slightly for large samples) but leaves the test of significance in equation (1.8) unchanged.

In the example, the significant R^2 and R_{adj}^2 for the regression of highest held degree on father's education, respondent's degree aspiration, and college selectivity are 0.108 and 0.107, respectively. Approximately 11% of the variability in *Degree* can be attributed to the three independent variables *FaEd*, *DegreAsp*, and *Selctvty*. This relatively small percentage of explained variability may be interpreted as evidence that the observed data do not adequately fit the hypothesized regression model. Other independent variables that were not included in the analysis could be considered as possibly better predictors of *Degree* than those used here. Note, however, the increase in R^2 when comparing the coefficient of determination from the multiple regression to the one from the simple regression. Since the latter is a submodel with respect to the full multiple regression, the difference in R^2s can be statistically tested by the F-ratio

$$F = \frac{(R_{full}^2 - R_{sub}^2)/(NX_{full} - NX_{sub})}{(1 - R_{full}^2)/(n - NX_{full} - 1)}, \qquad (1.17)$$

with $df_{num} = (NX_{full} - NX_{sub})$ and $df_{den} = (n - NX_{full} - 1)$. Using results from the multiple and simple regression examples, equation (1.17) becomes

$$F = \frac{(0.108 - 0.017)/(3 - 1)}{(1 - 0.108)/(3094 - 3 - 1)} = 157.618$$

with $df_{num} = 2$ and $df_{den} = 3090$. Thus, even though the coefficient of determination for the multiple regression is relatively small, the inclusion of two additional predictor variables (respondent's degree aspiration and college selectivity) in the simple regression equation significantly increases the proportion of explained variability in highest held academic degree.

Table 1.5. LISREL Input File for a Multiple
Linear Regression Example

1	Example 1.2. Multiple Linear Regression
2	DA NI = 4 NO = 3094
3	LA
4	Degree FaEd DegreAsp Selctvty
5	CM SY
6	.925
7	.188 2.283
8	.247 .187 1.028
9	.486 .902 .432 3.960
10	ME
11	4.535 3.747 4.003 5.016
12	MO NY = 1 NX = 3 AL = FR
13	OU SC ND = 3

LISREL EXAMPLE 1.2: MULTIPLE LINEAR REGRESSION OF ACADEMIC DEGREE ON FATHER'S EDUCATIONAL LEVEL, DEGREE ASPIRATIONS, AND COLLEGE SELECTIVITY. A LISREL analysis of the multiple linear regression example of respondent's highest held degree (*Degree*) on father's educational level (*FaEd*), respondent's degree aspirations (*DegreAsp*), and college selectivity (*Selctvty*) is presented below. As before, the input file will be explained, and selected output will verify that LISREL's maximum likelihood parameter estimates are equal to those obtained from an OLS regression. Consult Appendix A for a LISREL analysis of the example using the SIMPLIS command language.

LISREL Input. The input program is shown in Table 1.5. An explanation of changes from the LISREL program for the simple regression example (Table 1.1) is as follows:

Lines 1 and 2: Title and DAta lines. The first program line (optional) indicates a new title and, since the multiple regression example based on n = 3094 male respondents includes four variables (*Degree, FaEd, DegreAsp*, and *Selctvty*), NI = 4 and NO = 3094 is specified on the (required) DA line.

Lines 3 and 4: LAbel lines (optional). The input variables are given descriptive names consisting of no more than eight characters.

Lines 5 through 9: Data input lines (required). In this example, CM SY on line 5 indicates that a Symmetric variance/Covariance Matrix—instead of a correlation matrix as in the previous example (Table 1.1)—of the variables *Degree, FaEd, DegreAsp*, and *Selctvty* is used as data input.

Lines 10 and 11: MEans line (conditional). For illustrative purposes, the intercept term of the regression equation is estimated (line 12); thus, the sample means of the four variables *Degree, FaEd, DegreAsp*, and *Selctvty* are

Table 1.6. Selected LISREL Output of a Multiple Linear
Regression Example

LISREL ESTIMATES (MAXIMUM LIKELIHOOD)
GAMMA

	FaEd	DegreAsp	Selctvty
Degree	0.029	0.195	0.095
	(0.011)	(0.017)	(0.009)
	2.543	11.804	10.823

SQUARED MULTIPLE CORRELATIONS FOR STRUCTURAL EQUATIONS

Degree
0.108

ALPHA

Degree
3.170
(0.077)
41.288

STANDARDIZED SOLUTION
GAMMA

	FaEd	DegreAsp	Selctvty
Degree	0.045	0.206	0.196

specified. Note that a Standard Deviation (SD) line is not required since a variance/covariance matrix is used as input.

Lines 12 and 13: MOdel and OUtput lines (required). For the multiple regression example, *Degree* still is the only dependent variable (NY = 1), but now three independent variables (*FaEd*, *DegreAsp*, and *Selctvty*) are involved in the analysis; thus, NY = 1 and NX = 3 is specified on the MO line. The statement AL = FR indicates that the intercept parameter α is to be estimated. Finally, the OUtput line includes the same options that were explained in Example 1.1 [Standardized Completely (SC) and three-decimal accuracy of results (ND = 3)).

Selected LISREL Output. Table 1.6 contains selected output from the multiple linear regression analysis of *Degree* on *FaEd*, *DegreAsp*, and *Selctvty*. LISREL parameter estimates ($\hat{\gamma}_{Y1} = 0.029$, $\hat{\gamma}_{Y2} = 0.195$, $\hat{\gamma}_{Y3} = 0.095$, and $\hat{\alpha} = 3.170$) and the coefficient of determination ($R^2 = 0.108$) are equal to the OLS estimates previously discussed.

All t-values in Table 1.6 are greater than 2 and indicate that the estimated coefficients are significantly different from 0. In particular, all three independent variables, *FaEd*, *DegreAsp*, and *Selctvty*, can be interpreted as being significant predictors of the dependent variable, *Degree*. Finally, standard-

Table 1.7. EQS Input File for a Multiple Linear Regression Example

1	/TITLE
2	Example 1.2. Multiple Linear Regression
3	/SPECIFICATIONS
4	Cases = 3094; Variables = 4;
5	/LABELS
6	V1 = FaEd; V2 = DegreAsp; V3 = Selctvty; V4 = Degree;
7	/EQUATIONS
8	V4 = *V1 + *V2 + *V3 + E4;
9	/VARIANCES
10	V1 = *; V2 = *; V3 = *; E4 = *;
11	/COVARIANCES
12	V2,V1 = *; V3,V1 = *; V3,V2 = *;
13	/MATRIX
14	2.283
15	.187 1.028
16	.902 .432 3.960
17	.188 .247 .486 .925
18	/END

ized estimates of regression coefficient are calculated as $\hat{\gamma}_{(\text{stand})Y1} = 0.045$, $\hat{\gamma}_{(\text{stand})Y2} = 0.206$, $\hat{\gamma}_{(\text{stand})Y3} = 0.196$, which, again, are equal to the corresponding OLS estimates discussed earlier.

EQS EXAMPLE 1.2: MULTIPLE LINEAR REGRESSION OF ACADEMIC DEGREE ON FATHER'S EDUCATIONAL LEVEL, DEGREE ASPIRATIONS, AND COLLEGE SELECTIVITY

EQS Input. Table 1.7 lists an EQS program for the multiple regression example (*Degree* on *FaEd*, *DegreAsp*, and *Selectvty*). In the following we explain only noteworthy changes to the input file in Table 1.3. Recall that all information within sections of an EQS program must end with a semicolon (;), except within the sections related to data input near the end of the program.

Lines 1 through 4: /TITLE and /SPECIFICATIONS sections (optional and required, respectively). Line 2 assigns a new title to the program as part of the /TITLE section. Since a covariance matrix is used as input (lines 14 through 17), the statement "Matrix = Covariance;" could have been specified in the /SPECIFICATIONS section on line 4. However, the specification of a covariance matrix as input is the default in EQS and does not need to be mentioned.

Lines 5 and 6: /LABELS section (optional). Following the keyword /LABELS, descriptive names are given to the input variables V1 (*FaEd*), V2 (*DegreAsp*), V3 (*Selctvty*), and V4 (*Degree*). Recall that, according to the

Bentler-Weeks notation, Variables are denoted by the letter V and are numbered according to their order in the input matrix (lines 14 through 17).

Lines 7 through 10: /EQUATIONS and /VARIANCES sections (both required). On line 8, the multiple linear regression equation is specified without the intercept term; see the discussion in Example 1.1. The regression coefficients to be estimated are indicated by asterisks (*) directly in front of the associated independent variables [V1 (*FaEd*), V2 (*DegreAsp*), and V3 (*Selctvty*)]. The error term, E4, associated with the dependent variable, *Degree*, also is considered an independent variable in the context of the Bentler-Weeks model. However, its associated regression coefficient [the implied constant 1.0 in equation (1.9)] does not need to be estimated; hence, there is no asterisk in front of E4. Finally, line 10 identifies the variances of all four independent variables that need to be specified as unknown parameters to be estimated by EQS.

Lines 11 and 12: /COVARIANCES section (conditional). Hypothesized nonzero covariances among independent variables also must be specified in EQS (the reason will become apparent during the discussion of path analysis). There are four independent variables in the current multiple regression model (V1, V2, V3, and E4). Under OLS assumptions, however, error terms do not covary with observed independent variables. Thus, only three nonzero covariances ($\sigma_{V2,V1}$, $\sigma_{V3,V1}$, and $\sigma_{V3,V2}$) need to be estimated, which is specified by the three EQS statements "V2,V1 = *;" "V3,V1 = *;" "V3,V2 = *;" on line 12.

Lines 13 through 18: /MATRIX and /END program sections (conditional and required, respectively). Following the /MATRIX command, program lines 14 through 17 specify a lower triangular matrix containing the variances and covariances of the input variables *FaEd, DegreAsp, Selctvty*, and *Degree*. The program ends with the /END keyword.

Selected EQS Output. The output for the multiple regression of *Degree* (V4) on *FaEd* (V1), *DegreAsp* (V2), and *Selctvty* (V3) is reproduced partially in Table 1.8 (estimates of variances and covariances of independent variables

Table 1.8. Selected EQS Output of a Multiple Linear Regression Example

---

MAXIMUM LIKELIHOOD SOLUTION (NORMAL DISTRIBUTION THEORY)

MEASUREMENT EQUATIONS WITH STANDARD ERRORS AND TEST STATISTICS

Degree = V4 =	.029*V1 +	.195*V2 +	.095*V3 +	1.000 E4
	.011	.017	.009	
	2.544	11.810	10.829	

STANDARDIZED SOLUTION:

Degree = V4 =	.045*V1 +	.206*V2 +	.196*V3 +	.945 E4

---

are not shown). As in Example 1.1, the coefficient of determination R^2 can be computed from the standardized coefficient associated with E4 [the error term for *Degree* (V4)]; that is, $R^2 = 1 - 0.945^2 = 0.107$. All EQS results in Table 1.8 correspond within rounding error to the LISREL results presented in Table 1.6.

Before starting the discussion of classical path analysis, a potentially serious problem in the interpretation of multiple regression results must be mentioned. In general, the researcher attempts to identify a set of independent variables with members that highly correlate with the dependent variable but that have low correlations among themselves. In its simplest form, the problem of *multicolinearity* can develop when one or more independent variables are highly correlated with one or more of the other independent variables. In the extreme case when $R^2 = 1$ for some independent variables, there exists no unique solution for the regression coefficients. For instance, suppose that $\hat{\rho}^2_{X_2 X_1} = 1$ in a regression of a dependent variable, Y, on two independent variables, X_1 and X_2. Then essentially we have that $X_1 = X_2$, so that

$$\hat{Y} = \hat{a} + \hat{\gamma}_{Y1} X_1 + \hat{\gamma}_{Y2} X_2$$

$$= \hat{a} + (\hat{\gamma}_{Y1} + \hat{\gamma}_{Y2}) X_1,$$

which implies that $\hat{\gamma}_{Y1} + \hat{\gamma}_{Y2} = c$, for some constant c. Thus, unique values of $\hat{\gamma}_{Y1}$ and $\hat{\gamma}_{Y2}$ cannot be calculated; rather, there exists an infinite number of possible values for the regression coefficients that satisfy the equation $\hat{\gamma}_{Y1} + \hat{\gamma}_{Y2} = c$. The coefficients are said to be *underidentified* since there is not enough information to uniquely estimate them.

When $R^2 < 1$, a unique solution exists, but, if independent variables are highly correlated, estimates of regression coefficients can become very unstable (large changes in coefficient estimates result from only small changes in observed correlations or covariances), a situation referred to as *empirical underidentification*. As a general rule of thumb, multicolinearity *might* be present if:

1. absolute values of one or more of the zero-order correlation coefficients between independent variables are relatively high, say 0.70 or larger;
2. one or more of the metric or standardized regression coefficients have theory contradicting signs;
3. one or more of the standardized regression weights are very large;
4. one or more of the standard errors of regression coefficients are unusually large; or if
5. the regression equation has a large overall R^2 but few (if any) significant independent variables.

Results from the above LISREL and EQS regression analyses of *Degree* on *FaEd*, *DegreeAsp*, and *Selctvty* indicate that multicolinearity does not need to be a concern in this example: Absolute values of correlations among the independent variables are 0.3 or less (see Appendix D); estimated regression coefficients are positive as expected; all standardized coefficients are less than 1.0; standard errors associated with the regression coefficients are relatively small; and significance of the regression coefficients is consistent with the significance of the R^2.

Finally, when multicolinearity *is* present in a regression analysis, the researcher could (a) delete the offending variable(s) from the regression equation, (b) combine the involved variables into a single variable, or (c) use the highly correlated independent variables as indicators of one underlying construct, as is discussed in the next chapter on confirmatory factor analysis.

Classical Path Analysis

Wright's (1921, 1934) path analysis did not find its way into the social science literature until Duncan (1966) introduced the method in sociology. Since then, many applications have appeared in the research journals of most behavioral sciences. Although based on *correlational* data, path analysis provides the researcher with a multivariate (more than one dependent variable) method to estimate *structurally* interpretable terms—the direct, indirect, and total effects among a set of variables—provided a correct a priori path model, i.e., a theory-derived structure of the involved variables, is specified. Path diagrams are graphical representations of these a priori specified structures, and, if appropriate assumptions are met, ordinary least squares (OLS) estimates of regression coefficients can be used to estimate the strengths of the structural relationships specified in the diagram.

Path Diagrams and Structural Equations

The foregoing discussions and analyses of the influence of parents' socio-economic status (SES), respondent's motivation, and college prestige on respondent's SES can be extended by imposing an explicit and more detailed structure on these four constructs. A path diagram depicting one possible structure among the variables is presented in Figure 1.1: (a) Parents' SES is hypothesized to influence (i) respondent's motivation, (ii) the prestige of the colleges that respondents attended, and (iii) respondent's SES; (b) motivation is thought to effect (i) college prestige and (ii) respondent's SES; and (c) college prestige is hypothesized to influence respondent's SES. As before, father's educational level (*FaEd*) might be used as an indicator variable for parents' SES, respondent's degree aspirations (*DegreAsp*) can serve as a measure of motivation, selectivity (*Selctvty*) gives some indication of college prestige, and highest held academic degree (*Degree*) could be used as a measure of respondent's SES.

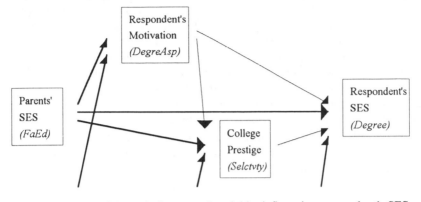

Figure 1.1. A possible path diagram of variables influencing respondent's SES.

In general, *path diagrams* are used to graphically display the a priori hypothesized structure among the variables in a model. For any two variables, X and Y, four basic relationships are possible:

1. $X \rightarrow Y$ X might structurally influence Y, but not vice versa.
2. $X \leftarrow Y$ Y might structurally influence X, but not vice versa.
3. $X \rightleftarrows Y$ X might structurally influence Y, and Y might structurally influence X.
4. X Y No structural relation is hypothesized between X and Y, but the variables might covary.

Structures that include relationships 1, 2, and/or 4 are called *recursive* models and are the focus in this text; models that involve variables hypothesized to structurally influence each other (relationship 3) are called *nonrecursive*, and their analysis is not discussed in this book (but see, for example, Berry, 1984). Unidirectional straight arrows in a path diagram represent structural influences from one variable to another, while bidirectional curved arrows represent the covariation between variables with *no* structural relationship specified or analyzed. Arrows pointing to variables from outside the model represent the collection of all other unmeasured influences, which usually are referred to as errors or disturbances. Independent variables that are exclusively influenced by factors lying outside the model are called *exogenous* (denoted as $X_j; j = 1, \ldots, NX$), while variables that are hypothesized to be influenced from inside the model are called *endogenous* (denoted as $Y_i; i = 1, \ldots, NY$). In Figure 1.1, parents' SES ($X_1 = Faed$) is the only exogenous variable, while motivation ($Y_1 = DegreAsp$), college prestige ($Y_2 = Selctvty$), and respondent's SES ($Y_3 = Degree$) are all endogenous variables. Following usual convention, structural flow among variables in any graphically presented structural equation model is shown from left to right and/or from top to bottom in a path diagram; i.e., endogenous variables are drawn to the right of and/or are shown below hypothesized structurally antecedent vari-

ables. Finally, observed variables in a path diagram are enclosed by squares
or rectangles (as in Figure 1.1), while latent, unobserved variables are shown
in circles or ellipses (see Chapter 2).

A path diagram represents a set of *structural equations*, that is, a set of
univariate regression equations with specified structure among the variables
in the model. *Structural coefficients* are the regression coefficients in these
equations and indicate two types of hypothesized relationships: (1) the structural effects of endogenous on other endogenous variables, denoted by β
(*beta*); and (2) the structural effects of exogenous on endogenous variables,
denoted by γ (*gamma*). The recursive model in Figure 1.1 represents the
following three structural equation:

$$Y_1 = \alpha_1 + \gamma_{11} X_1 + \zeta_1, \tag{1.18}$$

$$Y_2 = \alpha_2 + \beta_{21} Y_1 + \gamma_{21} X_1 + \zeta_2, \tag{1.19}$$

$$Y_3 = \alpha_3 + \beta_{31} Y_1 + \beta_{32} Y_2 + \gamma_{31} X_1 + \zeta_3, \tag{1.20}$$

where α_i (*alpha*) denotes the intercept term associated with each endogenous
variable (Y_i), $\beta_{ii'}$ (*beta*) is the *path* or *structural coefficient* in the regression of
endogenous (Y_i) on other endogenous ($Y_{i'}$) variables, γ_{ij} (*gamma*) denotes the
structural coefficient in the regression of endogenous (Y_i) on exogenous (X_j)
variables, ζ_i is the error term associated with an endogenous variable, Y_i, and
i and j are subscripts denoting the endogenous and exogenous variables,
respectively.

Figure 1.2 shows the same model as Figure 1.1 with general notation and
all structural coefficients except the intercept terms.

The set of structural equations (1.18)–(1.20) can be represented conve-
niently by the matrix equation

$$\begin{bmatrix} Y_1 \\ Y_2 \\ Y_3 \end{bmatrix} = \begin{bmatrix} \alpha_1 \\ \alpha_2 \\ \alpha_3 \end{bmatrix} + \begin{bmatrix} 0 & 0 & 0 \\ \beta_{21} & 0 & 0 \\ \beta_{31} & \beta_{32} & 0 \end{bmatrix} \begin{bmatrix} Y_1 \\ Y_2 \\ Y_3 \end{bmatrix} + \begin{bmatrix} \gamma_{11} \\ \gamma_{21} \\ \gamma_{31} \end{bmatrix} [X_1] + \begin{bmatrix} \zeta_1 \\ \zeta_2 \\ \zeta_3 \end{bmatrix}. \tag{1.21}$$

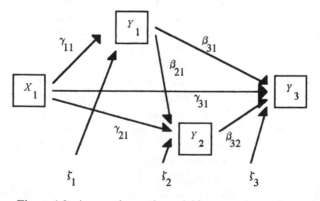

Figure 1.2. A recursive path model in general notation.

This representation is in the form of a general path analytical model,

$$\mathbf{Y} = \boldsymbol{\alpha} + \mathbf{BY} + \boldsymbol{\Gamma}\mathbf{X} + \boldsymbol{\zeta}, \qquad (1.22)$$

where $\mathbf{Y}$ is a $(NY \times 1)$ column vector of endogenous variables, $\mathbf{X}$ denotes a $(NX \times 1)$ column vector of exogenous variables, $\boldsymbol{\alpha}$ is a $(NY \times 1)$ column vector of intercept terms, $\mathbf{B}$ (**Beta** or **BE**) is a $(NY \times NY)$ matrix of structural coefficients from endogenous to other endogenous variables, $\boldsymbol{\Gamma}$ (**Gamma** or **GA**) is a $(NY \times NX)$ matrix of structural coefficients from exogenous to endogenous variables, $\boldsymbol{\zeta}$ (**Zeta** or **ZE**) denotes a $(NY \times 1)$ column vector of error terms of endogenous variables, NX is the number of exogenous variables, X, and NY is the number of endogenous variables, Y.

The nonzero elements in the two coefficient matrices $\mathbf{B}$ and $\boldsymbol{\Gamma}$ determine the structure of a particular path analytical model. In addition, a $(NX \times NX)$ variance/covariance matrix, $\boldsymbol{\Phi}$ (**Phi** or **PH**), of exogenous variables and a $(NY \times NY)$ variance/covariance matrix, $\boldsymbol{\Psi}$ (**Psi** or **PS**), of elements in $\boldsymbol{\zeta}$ can be specified. Taken together, the forms of these four basic matrices ($\mathbf{B} = \mathbf{BE}$, $\boldsymbol{\Gamma} = \mathbf{GA}$, $\boldsymbol{\Phi} = \mathbf{PH}$, and $\boldsymbol{\Psi} = \mathbf{PS}$) completely determine a particular path model. For the path diagram in Figure 1.2, the four matrices are

$$\mathbf{B} = \begin{bmatrix} 0 & 0 & 0 \\ \beta_{21} & 0 & 0 \\ \beta_{31} & \beta_{32} & 0 \end{bmatrix}, \quad \boldsymbol{\Gamma} = \begin{bmatrix} \gamma_{11} \\ \gamma_{21} \\ \gamma_{31} \end{bmatrix}, \quad \boldsymbol{\Phi} = [\sigma_{X_1}^2], \quad \text{and} \quad \boldsymbol{\Psi} = \begin{bmatrix} \sigma_{\zeta_1}^2 & 0 & 0 \\ 0 & \sigma_{\zeta_2}^2 & 0 \\ 0 & 0 & \sigma_{\zeta_3}^2 \end{bmatrix}.$$

Before considering a LISREL and EQS analysis of the example in Figures 1.1 and 1.2, three comments must be made. First, many researchers using SEM techniques express the values of observed variables as deviations from respective means since variability remains unchanged after transforming raw scores to deviation scores (see Appendix B). *From now on, unless otherwise stated, the assumption is made that variables are expressed as deviation scores.* The advantage of this convention is that the intercept terms in $\boldsymbol{\alpha}$ in equation (1.22) become 0, reducing the matrix form of a general path analysis model to

$$\mathbf{Y} = \mathbf{BY} + \boldsymbol{\Gamma}\mathbf{X} + \boldsymbol{\zeta}. \qquad (1.23)$$

It is, however, possible to extend the methods presented here to models that require the estimation of intercept terms (see, Bentler, 1993, Chapter 8; and Jöreskog and Sörbom, 1993a, Chapter 10).

Second, within a classical path analysis framework, some statistical assumptions need to be accepted to prevent problems with parameter estimation and interpretation. Specifically, for the path analysis models discussed here, it is assumed that:

1. The exogenous and endogenous variables are measured with no or negligible error and have a mean of 0 $[E(\mathbf{X}) = E(\mathbf{Y}) = \mathbf{0}]$.
2. The structural relations from the exogenous to the endogenous observed variables are linear.

3. The error terms in ζ in equation (1.22) (a) have a mean of 0 $[E(\zeta) = \mathbf{0}]$ and a constant variance across observations; (b) are independent, i.e., uncorrelated across observations; (c) are uncorrelated with the exogenous variables $[E(\mathbf{X}\zeta') = E(\zeta\mathbf{X}') = \mathbf{0}]$; and (d) are uncorrelated across equations, that is, the matrix $\mathbf{\Psi}$ (variance/covariance matrix of elements in ζ) has no nonzero off-diagonal elements.

Note, however, that for the more general structural equation models presented in Chapter 3, these restrictions can be relaxed so that (a) measurement errors of exogenous and endogenous variables can be incorporated into a model, and (b) disturbances are allowed to covary across equations, that is, the matrix $\mathbf{\Psi}$ may have nonzero off-diagonal elements.

Third, the minimum sample size required for SEM analyses must be considered. Bentler (1993) recommended that the ratio of sample size to the number of parameters to be estimated in a model is at the very least $5:1$, preferably much larger, say $10:1$ or $50:1$, if statistical significance tests are to be trusted. For the models in this chapter, all based on $n = 3094$ respondents, this rule of thumb clearly is satisfied; however, some illustrations in later chapters should be interpreted with some caution since they are based on $n = 167$ respondents only.

For future reference, Figure 1.3 schematically presents the notation, order, and role of the four basic matrices ($\mathbf{B}$, $\mathbf{\Gamma}$, $\mathbf{\Phi}$, and $\mathbf{\Psi}$) in path analytical models. Only two exogenous and two endogenous variables that are *minimally* connected are shown to keep the figure concise.

LISREL Example 1.3: A Path Analysis of the Effects of Father's Educational Level on Respondent's Academic Degree. The parameters in the two matrices $\mathbf{B}$ and $\mathbf{\Gamma}$ in equations (1.22) and (1.23) that represent the structure in the path analysis example in Figures 1.1 and 1.2 might be estimated with any standard OLS regression program. After all, the figures simply represent three structural equations that statistically are no different from univariate multiple linear regression equations *if* OLS assumptions are met. As will be apparent soon, however, there are several advantages of using a software package such as LISREL or EQS to estimate structural coefficients in path analytical models. The next two sections present the LISREL input program and selected output for the estimation of the population parameters in the four matrices $\mathbf{B}$, $\mathbf{\Gamma}$, $\mathbf{\Phi}$, and $\mathbf{\Psi}$ for the model depicted in Figure 1.2. In Appendix A, a LISREL analysis of this model is presented using the SIMPLIS command language, which does not require the specification of the four basic matrices.

LISREL Input. An input program for the estimation of the parameters of the path model in Figure 1.2 is shown in Table 1.9.

Lines 1 through 9: Title (optional), DAta (required), LAbel (optional), and data input (required) lines. The first nine lines of the input do not differ from

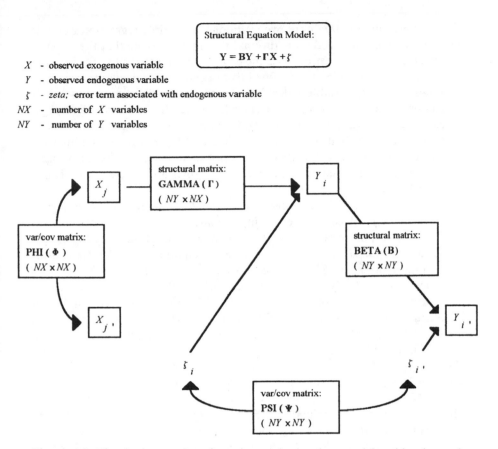

Figure 1.3. The basic matrices for structural equation models with observed variables.

Table 1.9. LISREL Input File for the Model in Figure 1.2

1	Example 1.3. Path Analysis With one Exogenous Variable
2	DA NI = 4 NO = 3094
3	LA
4	Degree FaEd DegreAsp Selctvty
5	CM SY
6	.925
7	.188 2.283
8	.247 .187 1.028
9	.486 .902 .432 3.960
10	SE
11	DegreAsp Selctvty Degree FaEd
12	MO NY = 3 NX = 1 BE = SD GA = FU PH = SY PS = DI
13	OU SC ND = 3

the ones presented in the input program for the multiple regression example (Table 1.5) and, thus, need no additional explanation. Note that the input of sample means is not required since the estimation of intercept terms is not of interest, that is, variables are assumed to be expressed as deviation scores.

Lines 10 and 11: Variable SElect lines (conditional). This line allows for the SElection and/or reordering of variables from the input matrix. All endogenous (dependent) variables, *Y*, must be listed in structural order (from left to right and/or top to bottom in the path diagram) *before* the exogenous variables, *X*. Variables can be listed either by their labels or by their place number in the input matrix. If not all NI variables are selected, the SE line must end with a slash (/).

On the *SE* line for the example in Figure 1.2, the three endogenous variables (*DegreAsp*, place number in input matrix = 3; *Selctvty*, place number = 4; *Degree*, place number = 1) are selected first, followed by the exogenous variable (*FaEd*, place number = 2). Since all input variables are considered in the analysis, the SE line does not need to end in a slash (/).

Lines 12 and 13: MOdel and OUtput lines (both required). The specific structure of a path model is given on the MO line. First, the number of endogenous and exogenous variables is given (NY and NX, respectively), followed by the specification of the four basic matrices $\mathbf{B} = \mathbf{BE}$, $\mathbf{\Gamma} = \mathbf{GA}$, $\mathbf{\Phi} = \mathbf{PH}$, and $\mathbf{\Psi} = \mathbf{PS}$. In the example, NY = 3 (*DegreAsp*, *Selctvty*, and *Degree*) and NX = 1 (*FaEd*). As Figure 1.2 indicates, the estimates of coefficients β_{21}, β_{31}, and β_{32} among endogenous variables need to be calculated. Thus, $\mathbf{BE}$ is a (3 × 3) SubDiagonal (SD) matrix in which the elements below the main diagonal are estimated (BE = SD). Furthermore, since the exogenous variable *FaEd* (X_1) is hypothesized to influence all endogenous variables (Y_1, Y_2, and Y_3), all parameter estimates in the (3 × 1) $\mathbf{GA}$ matrix (γ_{11}, γ_{21}, and γ_{31}) need to be calculated. GA = FU indicates that $\mathbf{GA}$ is a FUll— in contrast to a subdiagonal or symmetric—matrix and that all structural coefficients from exogenous to endogenous variables are being estimated.

Since there is only one exogenous variable ($X_1 = FaEd$), the SYmmetric matrix $\mathbf{PH}$ is of order (1 × 1), containing only the variance of X_1. The statement "PH = SY" is LISREL's default for path analytical models since the program assumes that all variances and covariances of exogenous observed variables are being estimated. Also, the elements of ζ (column vector of error terms of endogenous variables) are hypothesized not to covary. The variance/covariance matrix of error terms, $\mathbf{PS}$, therefore is specified to be DIagonal (PS = DI), indicating that only the variances of error terms are estimated. Finally, the required OUtput line includes the same options that were used in the regression examples 1.1 and 1.2.

Selected LISREL Output. Table 1.10 contains selected output based on the input program in Table 1.9. LISREL's maximum likelihood (ML) parameter estimates of elements in $\mathbf{B}$, $\mathbf{\Gamma}$, $\mathbf{\Phi}$, and $\mathbf{\Psi}$ are given (with their respective

Table 1.10. Selected LISREL Output from the Analysis of the Model in
Figure 1.2

LISREL ESTIMATES (MAXIMUM LIKELIHOOD)

BETA

	DegreAsp	Selctvty	Degree
DegreAsp	—	—	—
Selctvty	0.354 (0.033) 10.612	—	—
Degree	0.195 (0.017) 11.808	0.095 (0.009) 10.827	—

GAMMA

	FaEd
DegreAsp	0.082 (0.012) 6.839
Selctvty	0.366 (0.022) 16.374
Degree	0.029 (0.011) 2.543

PHI

FaEd
2.283

PSI

DegreAsp	Selctvty	Degree
1.013 (0.026) 39.319	3.477 (0.088) 39.319	0.825 (0.021) 39.319

SQUARED MULTIPLE CORRELATIONS FOR STRUCTURAL EQUATIONS

DegreAsp	Selctvty	Degree
0.015	0.122	0.108

STANDARDIZED SOLUTION

BETA

	DegreAsp	Selctvty	Degree
DegreAsp	—	—	—
Selctvty	0.180	—	—
Degree	0.206	0.196	—

Table 1.10 (*cont.*)

	GAMMA			
		FaEd		
DegreAsp		0.122		
Selctvty		0.278		
Degree		0.045		
	PSI			
		DegreAsp	Selctvty	Degree
		0.985	0.878	0.892

standard errors and t-values) followed by the R^2-values and the standardized solutions.

Using the output in Table 1.10, the three unstandardized structural equations can be estimated by

$$\hat{Y}_1 = \hat{\gamma}_{11}X_1 = 0.082X_1, \tag{1.24}$$

$$\hat{Y}_2 = \hat{\beta}_{21}Y_1 + \hat{\gamma}_{21}X_1 = 0.354Y_1 + 0.366X_1, \tag{1.25}$$

$$\hat{Y}_3 = \hat{\beta}_{31}Y_1 + \hat{\beta}_{32}Y_2 + \hat{\gamma}_{31}X_1 = 0.195Y_1 + 0.095Y_2 + 0.029X_1. \tag{1.26}$$

Interpreting the printed coefficients of determination for the three structural equations, we conclude that only about 2% ($R^2_{Y_1} = 0.015$) of the variability in respondent's degree aspirations can be attributed to the variability in father's educational level; approximately 12% ($R^2_{Y_2} = 0.122$) of the variance in college selectivity is accounted for by the variability of the two variables that are hypothesized to affect it (*FaEd* and *DegreAsp*); and about 11% ($R^2_{Y_3} = 0.108$) of the variability in respondent's highest held academic degree is explained by father's education, respondent's degree aspiration, and college selectivity.

Finally, considering the standardized solutions and t-values for the estimates in the **B** and **Γ** matrices, the structural equations can be graphically displayed by redrawing Figure 1.1, this time attaching the estimates of both metric and standardized structural coefficients and indicating their statistical significance (see Figure 1.4).

EQS EXAMPLE 1.3: A PATH ANALYSIS OF THE EFFECTS OF FATHER'S EDUCATIONAL LEVEL ON RESPONDENT'S ACADEMIC DEGREE. The recursive path model of Figures 1.1 and 1.2 is reproduced in Figure 1.5 with the Bentler-Weeks notation to clarify the different notational systems used in LISREL and EQS. Recall that, here, observed Variables (V) in EQS are numbered consecutively from left to right and top to bottom in the path diagram.

EQS Input. Table 1.11 shows an EQS program for the analysis of the model in Figure 1.5. Since EQS (like the SIMPLIS command language in LISREL) does not require the user to explicitly name and specify the basic matrices **B**,

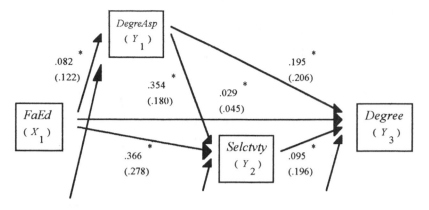

Figure 1.4. Redrawn path diagram of the model in Figure 1.1 including structural coefficients. *Note*: Standardized coefficients are enclosed in parentheses and coefficients that are at least twice their standard errors are indicated by *.

Table 1.11. EQS Input File for the Model in Figure 1.5

1	/TITLE
2	Example 1.3. Path Analysis With one Exogenous Variable
3	/SPECIFICATIONS
4	Cases = 3094; Variables = 4;
5	/LABELS
6	V1 = FaEd; V2 = DegreAsp; V3 = Selctvty; V4 = Degree;
7	/EQUATIONS
8	V2 = *V1 + E2;
9	V3 = *V1 + *V2 + E3;
10	V4 = *V1 + *V2 + *V3 + E4;
11	/VARIANCES
12	V1 = *; E2 = *; E3 = *; E4 = *;
13	/MATRIX
14	2.283
15	.187 1.028
16	.902 .432 3.960
17	.188 .247 .486 .925
18	/END

Γ, Φ, and Ψ, the reader already is familiar with most of the EQS syntax required to analyze the model in Figure 1.5. It suffices to discuss only two key sections of the input program (the /EQUATIONS and /VARIANCES sections) that *implicitly* specify the matrix elements that are being estimated.

Lines 7 through 10: /EQUATIONS section. The model in Figure 1.5 represents a system of three structural equations that must be specified in EQS. Lines 8, 9, and 10 of the input program correspond to equations (1.18), (1.19),

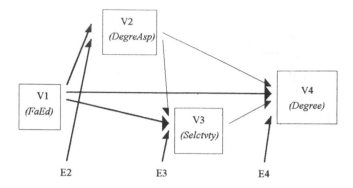

Figure 1.5. A recursive path model in Bentler-Weeks notation.

and (1.20), respectively. Asterisks indicate the structural coefficients to be estimated by EQS and correspond to the free parameters in the **B** and **Γ** matrices of equations (1.21) and (1.22).

Lines 11 and 12: /VARIANCES section. This section must contain all variances of independent variables (in the sense of Bentler-Weeks). In the model of Figure 1.5, there are four independent variables: V1 (*FaEd*) and E2, E3, and E4 [the error terms associated with variables V2 (*DegreAsp*), V3 (*Selctvty*), and V4 (*Degree*)]. Thus, their variances are specified as parameters to be estimated on line 12 of the input file: The first statement (V1 = *;) corresponds to the free parameter in the variance/covariance matrix **Φ**, while the remaining statements (E2 = *; E3 = *; and E4 = *;) on line 12 correspond to the parameters to be estimated in the **Ψ** matrix (consult Figure 1.3 for clarification).

Selected EQS Output. Parts of the output are shown in Table 1.12. Comparing the results to LISREL estimates of the elements in **B**, **Γ**, **Φ**, and **Ψ** (Table 1.10) illustrates the numerical equivalence of the different approaches taken in LISREL and EQS. As in previous EQS examples, the coefficient of determination, R^2, for each of the three structural equations can be computed from the standardized coefficients associated with the error terms that are given at the end of the EQS output. That is, $R^2_{V2} = 1 - 0.993^2 = 0.014$, $R^2_{V3} = 1 - 0.937^2 = 0.122$, and $R^2_{V4} = 1 - 0.945^2 = 0.107$ for the V2 (*DegreAsp*), V3 (*Selctvty*), and V4 (*Degree*) equations, respectively.

The Decomposition of Covariances and the Direct, Indirect, and Total Structural Effects

The underlying statistical assumptions, the equations, and the interpretations of coefficients in classical path analysis are similar to those in multiple regres-

Table 1.12. Selected EQS Output from the Analysis of the Model in Figure 1.5

MAXIMUM LIKELIHOOD SOLUTION (NORMAL DISTRIBUTION THEORY)

MEASUREMENT EQUATIONS WITH STANDARD ERRORS AND TEST STATISTICS

DegreAsp = V2 = .082*V1 +   1.000 E2
                 .012
                 6.840

Selctvty = V3 = .354*V2 +   .366*V1 +   1.000 E3
                 .033   .022
                 10.614   16.376

Degree = V4 =   .195*V2 +   .095*V3 +   .029*V1 +   1.000 E4
                 .017   0.009   .011
                 11.810   10.829   2.544

VARIANCES OF INDEPENDENT VARIABLES

V	F	E	D
V1 − FaEd	2.283* I	I E2 − DegreAsp	1.013* I
	.058 I	I	.026 I
	39.326 I	I	39.326 I
	I	I	I
	I	I E3 − Selctvty	3.477* I
	I	I	.088 I
	I	I	39.326 I
	I	I	I
	I	I E4 − Degree	.825* I
	I	I	0.21 I
	I	I	39.326 I
	I	I	I

STANDARDIZED SOLUTION:

DegreAsp = V2 = .122*V1 +   .993 E2
Selctvty = V3 = .180*V2 +   .278*V1 +   .937 E3
Degree = V4 =   .206*V2 +   .196*V3 +   .045*V1 +   .945 E4

---

sions, but structural relationships were specified a priori. One major advantage of path analysis is that, in addition to direct structural effects (DE), indirect effects (IE) through intervening variables can be estimated. For example, in Figure 1.4, father's educational level ($FaEd$) has a direct structural effect on $Degree$ (expressed by either the metric coefficient, $\hat{\gamma}_{31} = 0.029$, or the standardized weight, $\hat{\gamma}_{31(\text{stand})} = 0.045$); but $FaEd$ also has an indirect structural effect on $Degree$ through respondent's degree aspirations (represented by the product of the corresponding unstandardized or standardized coefficients, $\hat{\gamma}_{11}\hat{\beta}_{31} = 0.016$ or $\hat{\gamma}_{11(\text{stand})}\hat{\beta}_{31(\text{stand})} = 0.025$, respectively).

The computational definitions of a direct and indirect effect of an independent on a dependent variable in a general structural equation model are based on the fact that covariances between two variables can be completely decomposed and written as functions of the model-implied parameters. One way to accomplish this is through a generalized version of Kenny's (1979) *First Law of Path Analysis*:

$$\sigma_{YX} = \sum_q p_{Yq}\sigma_{qX}, \tag{1.27}$$

where q is a subscript denoting all variables (including error terms) with direct paths to variable Y and p_{Yq} denotes the path coefficient from variable q to variable Y. The law states that the covariance between two variables, X and Y, can be decomposed into the sum of products of structural coefficients of variables, q, with direct paths to Y and the covariances of these variables with X.

Consider the path diagram in Figure 1.2. Using equation (1.27), the covariance between Y_3 (*Degree*) and X_1 (*FaEd*) can be decomposed completely by following five steps:

1. Find all variables, q, that directly affect the endogenous variable Y_3. These are Y_1 (*DegreAsp*), Y_2 (*Selctvty*), X_1 (*FaEd*), and ζ_3, the error term associated with Y_3.
2. Record the structural coefficients that link the variables identified in Step 1 to the endogenous variable Y_3. These are β_{31}, β_{32}, γ_{31}, and 1.0, respectively.
3. Multiply each structural coefficient identified in Step 2 by the covariance of X_1 and the associated variable identified in Step 1; now sum the results. Thus,

$$\sigma_{Y_3X_1} = \beta_{31}\sigma_{Y_1X_1} + \beta_{32}\sigma_{Y_2X_1} + \gamma_{31}\sigma_{X_1X_1} + \sigma_{\zeta_3X_1}$$
$$= \beta_{31}\sigma_{Y_1X_1} + \beta_{32}\sigma_{Y_2X_1} + \gamma_{31}\sigma_{X_1}^2. \tag{1.28}$$

Note from Figure 1.2 that ζ_3 does not covary with X_1 ($\sigma_{\zeta_3X_1} = 0$). Hence, the right side of equation (1.28) does not involve the term $\sigma_{\zeta_3X_1}$.

4. The decomposition of $\sigma_{Y_3X_1}$ in equation (1.28) contains two covariances, $\sigma_{Y_1X_1}$ and $\sigma_{Y_2X_1}$, which in turn can be decomposed by repeating Steps 1 and 3 above. That is,

$$\sigma_{Y_1X_1} = \gamma_{11}\sigma_{X_1X_1} = \gamma_{11}\sigma_{X_1}^2, \tag{1.29}$$

and

$$\sigma_{Y_2X_1} = \beta_{21}\sigma_{Y_1X_1} + \gamma_{21}\sigma_{X_1X_1} = \beta_{21}(\gamma_{11}\sigma_{X_1}^2) + \gamma_{21}\sigma_{X_1}^2. \tag{1.30}$$

5. Finally, substitution of equations (1.29) and (1.30) into the right side of equation (1.28) yields

$$\sigma_{Y_3X_1} = \beta_{31}\gamma_{11}\sigma_{X_1}^2 + \beta_{32}\beta_{21}\gamma_{11}\sigma_{X_1}^2 + \beta_{32}\gamma_{21}\sigma_{X_1}^2 + \gamma_{31}\sigma_{X_1}^2. \tag{1.31}$$

Equation (1.31)—a complete decomposition of $\sigma_{Y_3 X_1}$ in terms of elements in the matrices $\mathbf{B}$, $\mathbf{\Gamma}$, and $\mathbf{\Phi}$—can be verified by numerical substitution from the LISREL estimates presented in Table 1.10 or Figure 1.4. That is,

$$\hat{\sigma}_{Y_3 X_1} = \hat{\beta}_{31}\hat{\gamma}_{11}\hat{\sigma}_{X_1}^2 + \hat{\beta}_{32}\hat{\beta}_{21}\hat{\gamma}_{11}\hat{\sigma}_{X_1}^2 + \hat{\beta}_{32}\hat{\gamma}_{21}\hat{\sigma}_{X_1}^2 + \hat{\gamma}_{31}\hat{\sigma}_{X_1}^2, \qquad (1.32)$$

which is numerically equivalent to

$$0.188 = 0.195(0.082)2.283 + 0.095(0.354)0.082(2.283)$$

$$+ 0.095(0.366)2.283 + 0.029(2.283).$$

When variables are standardized, their variances equal 1 and the decomposition simplifies to

$$\hat{\rho}_{Y_3 X_1} = \hat{\beta}_{31(\text{stand})}\hat{\gamma}_{11(\text{stand})} + \hat{\beta}_{32(\text{stand})}\hat{\beta}_{21(\text{stand})}\hat{\gamma}_{11(\text{stand})}$$

$$+ \hat{\beta}_{32(\text{stand})}\hat{\gamma}_{21(\text{stand})} + \hat{\gamma}_{31(\text{stand})}, \qquad (1.33)$$

which is equivalent to

$$0.129 = 0.206(0.122) + 0.196(0.180)0.122 + 0.196(0.278) + 0.045.$$

In equation (1.33), the product $\hat{\beta}_{31(\text{stand})}\hat{\gamma}_{11(\text{stand})} = 0.025$ is defined as the particular indirect structural effect of X_1 on Y_3 through the intervening variable Y_1; $\hat{\beta}_{32(\text{stand})}\hat{\beta}_{21(\text{stand})}\hat{\gamma}_{11(\text{stand})} = 0.004$ is the indirect effect of X_1 on Y_3 via the two intervening variables Y_1 and Y_2; $\hat{\beta}_{32(\text{stand})}\hat{\gamma}_{21(\text{stand})} = 0.054$ is the indirect effect via variable Y_2; and $\hat{\gamma}_{31(\text{stand})} = 0.045$ is the direct structural effect of X_1 on Y_3.

More generally, the *direct effect* (DE) of an independent on a dependent variable in a structural equation model is defined as the structural coefficient (metric or standardized) linking the two variables. A *particular indirect effect* of one variable on another through one or more intervening variables is defined by the product of associated structural coefficients (metric or standardized) that link the variables in the particular structural chain; the sum of *all* particular indirect effects is defined as the *total indirect effect* (IE). Finally, the sum of the direct effect (DE) and the indirect effect (IE) is defined as the *total effect* (TE) of an independent on a dependent variable (Alwin and Hauser, 1975; Duncan, 1975; also see Sobel, 1986).

Using the standardized coefficients shown in Figure 1.4, the direct, indirect, and total structural effects of *FaEd* on *Degree* can be computed as

$$DE_{Y_3 X_1} = \hat{\gamma}_{31(\text{stand})} = 0.045, \qquad (1.34)$$

$$IE_{Y_3 X_1} = \hat{\beta}_{31(\text{stand})}\hat{\gamma}_{11(\text{stand})} + \hat{\beta}_{32(\text{stand})}\hat{\beta}_{21(\text{stand})}\hat{\gamma}_{11(\text{stand})} + \hat{\beta}_{32(\text{stand})}\hat{\gamma}_{21(\text{stand})}$$

$$= 0.025 + 0.004 + 0.054 = 0.084, \qquad (1.35)$$

and

$$TE_{Y_3 X_1} = DE_{Y_2 X_1} + IE_{Y_3 X_1} = 0.045 + 0.084 = 0.129. \qquad (1.36)$$

The determination of structural effect components in larger, more general structural equation models can become a complex and cumbersome task since *all* possible combinations of paths between a particular pair of variables must be identified. Fortunately, SEM programs such as LISREL and EQS have incorporated matrix algorithms to determine the various effects (see Chapter 3 and Bollen, 1989; Fox, 1980; or Sobel, 1982, 1986, 1987). Here, it suffices to remember that

1. The direct effect (*DE*) of an endogenous on another endogenous variable is the coefficient in the **B** matrix that is associated with the two variables.
2. The direct effect of an exogenous on an endogenous variable is the structural coefficient in the **Γ** matrix that is associated with the two variables.
3. A particular indirect effect between two variables through specific intervening variables is computed by forming the product of structural coefficients in the **B** and/or **Γ** matrices along the particular path from the independent to the dependent variable.
4. The total indirect effect (*IE*) between two variables is defined as the sum of *all* particular indirect effects through possible intervening variables.
5. The total effect (*TE*) between two variables is defined as the sum of the direct and total indirect effect components.

LISREL EXAMPLE 1.4: THE DIRECT AND INDIRECT STRUCTURAL EFFECTS IN A PATH MODEL OF FATHER'S EDUCATIONAL LEVEL AND RESPONDENT'S HIGH SCHOOL RANK ON ACADEMIC DEGREE. An illustration of the estimation of direct, indirect, and total effects among variables in a path model is provided by adding a second exogenous variable—respondent's high school rank ($X_2 = HSRank$), a possible indicator variable of the more general construct of academic rank—to the model in Figure 1.1. The modified path diagram with the structural coefficients is presented in Figure 1.6.

Recalling that variables are assumed to be expressed as deviation scores,

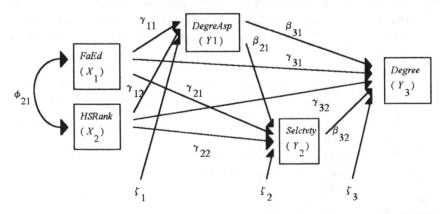

Figure 1.6. Another possible path diagram of variables influencing respondent's SES.

the structural equations represented by the model in Figure 1.6 can be written as

$$Y_1 = \gamma_{11}X_1 + \gamma_{12}X_2 + \zeta_1, \tag{1.37}$$

$$Y_2 = \beta_{21}Y_1 + \gamma_{21}X_1 + \gamma_{22}X_2 + \zeta_2, \tag{1.38}$$

$$Y_3 = \beta_{31}Y_1 + \beta_{32}Y_2 + \gamma_{31}X_1 + \gamma_{32}X_2 + \zeta_3; \tag{1.39}$$

while the four basic matrices, $\mathbf{B}$, $\mathbf{\Gamma}$, $\mathbf{\Phi}$, and $\mathbf{\Psi}$, now are

$$\mathbf{B} = \begin{bmatrix} 0 & 0 & 0 \\ \beta_{21} & 0 & 0 \\ \beta_{31} & \beta_{32} & 0 \end{bmatrix}, \quad \mathbf{\Gamma} = \begin{bmatrix} \gamma_{11} & \gamma_{12} \\ \gamma_{21} & \gamma_{22} \\ \gamma_{31} & \gamma_{32} \end{bmatrix},$$

$$\mathbf{\Phi} = \begin{bmatrix} \sigma_{X_1}^2 & \sigma_{X_1 X_2} \\ \sigma_{X_2 X_1} & \sigma_{X_2}^2 \end{bmatrix}, \quad \text{and} \quad \mathbf{\Psi} = \begin{bmatrix} \sigma_{\zeta_1}^2 & 0 & 0 \\ 0 & \sigma_{\zeta_2}^2 & 0 \\ 0 & 0 & \sigma_{\zeta_3}^2 \end{bmatrix}.$$

Note that the $\mathbf{\Phi}$ matrix contains a hypothesized non-zero covariance, $\sigma_{X_2 X_1}$, between the two exogenous variables *FaEd* and *HSRank* that is denoted by ϕ_{21} and is indicated by an associated bidirectional curved arrow in Figure 1.6.

The next two sections present the LISREL input and selected output for the estimation of the direct, indirect, and total structural effects among variables in the model shown in Figure 1.6.

LISREL Input. The program to estimate the model in Figure 1.6 is reproduced in Table 1.13. Since by now most of the input statements should be familiar, only the new LISREL options are discussed.

Table 1.13. LISREL Input File for the Model in Figure 1.6

1	Example 1.4. Path Analysis With Two Exogenous Variables
2	DA NI = 5 NO = 3094 MA = CM
3	LA
4	DegreAsp Selctvty Degree FaEd HSRank
5	KM SY
6	1
7	.214 1
8	.253 .254 1
9	.122 .300 .129 1
10	.194 .372 .189 .128 1
11	SD
12	1.014 1.990 .962 1.511 .777
13	MO NY = 3 NX = 2 BE = SD GA = FU PH = SY PS = DI
14	OU SC EF ND = 3

Line 2: Estimates of structural coefficients should be based on an analysis of a variance/covariance matrix irrespective of the form of the input data. Thus, the Matrix to be Analyzed (MA) is specified to be a Covariance Matrix (CM), the default in LISREL, even though the input matrix is a Symmetric [K]orrelation matrix (KM SY on line 5). For other permissible forms of MA, consult the LISREL manual.

Line 14: The option EF on the output line signals LISREL that all Effects (direct, indirect, and total) should be calculated. Since SC (Standardized Completely) also was requested, LISREL will print the unstandardized as well as the standardized effect components.

Selected LISREL Output. Table 1.14 contains the metric and standardized maximum likelihood estimates (equivalent to OLS estimates) of all parameters in the model depicted in Figure 1.6. The matrices **B** and **Γ** contain the direct structural effects among variables, and elements in additionally provided matrices are the indirect and total effects.

From the unstandardized results presented in Table 1.14, the three structural equations (1.37)–(1.39) for the model in Figure 1.6 can be estimated by

$$\hat{Y}_1 = \hat{\gamma}_{11}X_1 + \hat{\gamma}_{12}X_2$$
$$= 0.066X_1 + 0.237X_2, \tag{1.40}$$

$$\hat{Y}_2 = \hat{\beta}_{21}Y_1 + \hat{\gamma}_{21}X_1 + \hat{\gamma}_{22}X_2$$
$$= 0.241Y_1 + 0.322X_1 + 0.812X_2, \tag{1.41}$$

$$\hat{Y}_3 = \hat{\beta}_{31}Y_1 + \hat{\beta}_{32}Y_2 + \hat{\gamma}_{31}X_1 + \hat{\gamma}_{32}X_2$$
$$= 0.186Y_1 + 0.081Y_2 + 0.028X_1 + 0.103X_2. \tag{1.42}$$

Figure 1.7 shows the redrawn associated path diagram including metric and standardized structural coefficients and indicating their statistical significance (coefficients that are at least twice as large as their standard errors).

Coefficients of determination for the three structural equations indicate that: a) the exogenous variables *FaEd* and *HSRank* explain about 5% ($R^2_{Y_1} = 0.047$) of the variability in respondent's degree aspiration; b) variables *FaEd, HSRank*, and *DegreAsp* jointly explain 22% ($R^2_{Y_2} = 0.217$) of the variance in college selectivity; and c) about 11% ($R^2_{Y_3} = 0.114$) of the variability in respondent's highest degree can be attributed to variability in *FaEd, HSRank, DegreAsp*, and *Selctvty*.

Next, consider the matrices of total and indirect effects, presented in Table 1.14 first for exogenous on endogenous variables and then for endogenous on other endogenous variables. For example, the unstandardized indirect effect of respondent's high school rank on highest held academic degree through all intervening variables (*DegreAsp* and *Selctvty*) is estimated as 0.114 ($SE = 0.010$, $t = 11.163$). This can be verified from Figure 1.7 by identifying all possible structural chains between *HSRank* (X_2) and *Degree* (Y_3) that

Table 1.14. Selected LISREL Output from the Analysis of the Model in Figure 1.6

LISREL ESTIMATES (MAXIMUM LIKELIHOOD)

BETA

	DegreAsp	Selctvty	Degree
DegreAsp	—	—	—
Selctvty	0.241	—	—
	(0.032)		
	7.527		
Degree	0.186	0.081	—
	(0.017)	(0.009)	
	11.170	8.780	

GAMMA

	FaEd	HSRank
DegreAsp	0.066	0.237
	(0.012)	(0.023)
	5.580	10.245
Selctvty	0.322	0.812
	(0.021)	(0.042)
	15.162	19.427
Degree	0.028	0.103
	(0.011)	(0.023)
	2.480	4.507

PHI

	FaEd	HSRank
FaEd	2.283	
HSRank	0.150	0.604

PSI

	DegreAsp	Selctvty	Degree
	0.980	3.099	0.820
	(0.025)	(0.079)	(0.021)
	39.313	39.313	39.313

SQUARED MULTIPLE CORRELATIONS FOR STRUCTURAL EQUATIONS

	DegreAsp	Selctvty	Degree
	0.047	0.217	0.114

STANDARDIZED SOLUTION

BETA

	DegreAsp	Selctvty	Degree
DegreAsp	—	—	—
Selctvty	0.123	—	—
Degree	0.196	0.168	—

Table 1.14 (*cont.*)

	GAMMA		
		FaEd	HSRank
DegreAsp		0.099	0.181
Selctvty		0.244	0.317
Degree		0.044	0.083

	PSI		
		DegreAsp	Selctvty	Degree
		0.953	0.783	0.886

TOTAL AND INDIRECT EFFECTS

TOTAL EFFECTS OF X ON Y

	FaEd	HSRank
DegreAsp	0.066	0.237
	(0.012)	(0.023)
	5.580	10.245
Selctvty	0.338	0.869
	(0.021)	(0.041)
	15.849	20.949
Degree	0.068	0.217
	(0.011)	(0.022)
	6.018	9.904

INDIRECT EFFECTS OF X ON Y

	FaEd	HSRank
DegreAsp	—	—
Selctvty	0.016	0.057
	(0.004)	(0.009)
	4.483	6.066
Degree	0.040	0.114
	(0.004)	(0.010)
	9.132	11.163

TOTAL EFFECTS OF Y ON Y

	DegreAsp	Selctvty	Degree
DegreAsp	—	—	—
Selctvty	0.241	—	—
	(0.032)		
	7.527		
Degree	0.205	0.081	—
	(0.017)	(0.009)	
	12.308	8.780	

Table 1.14 (*cont.*)

	INDIRECT EFFECTS OF Y ON Y		
	DegreAsp	Selctvty	Degree
DegreAsp	—	—	—
Selctvty	—	—	—
Degree	0.020 (0.003) 5.714	—	—

	STANDARDIZED TOTAL EFFECTS OF X ON Y	
	FaEd	HSRank
DegreAsp	0.099	0.181
Selctvty	0.257	0.339
Degree	0.107	0.175

	STANDARDIZED INDIRECT EFFECTS OF X ON Y	
	FaEd	HSRank
DegreAsp	—	—
Selctvty	0.012	0.022
Degree	0.062	0.092

	STANDARDIZED TOTAL EFFECTS OF Y ON Y		
	DegreAsp	Selctvty	Degree
DegreAsp	—	—	—
Selctvty	0.123	—	—
Degree	0.216	0.168	—

	STANDARDIZED INDIRECT EFFECTS OF Y ON Y		
	DegreAsp	Selctvty	Degree
DegreAsp	—	—	—
Selctvty	—	—	—
Degree	0.021	—	—

involve intervening variables, forming the products of unstandardized structural coefficients along each such chain, and summing the results. That is,

$$IE_{Y_3 X_2} = \hat{\beta}_{31} \hat{\gamma}_{12} + \hat{\beta}_{32} \hat{\beta}_{21} \hat{\gamma}_{12} + \hat{\beta}_{32} \hat{\gamma}_{22}$$

$$= 0.186(0.237) + 0.081(0.241)0.237 + 0.081(0.812) \quad (1.43)$$

$$= 0.044 + 0.005 + 0.066 = 0.114.$$

Also, the direct effect of *HSRank* on *Degree* is given at the beginning of the LISREL output in the estimated Γ matrix as

$$DE_{Y_3 X_2} = \hat{\gamma}_{32} = 0.103, \quad (1.44)$$

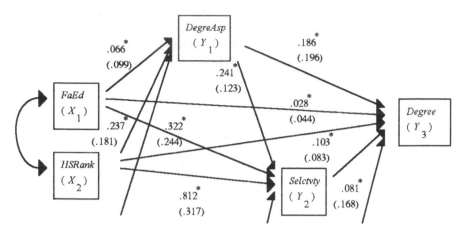

Figure 1.7. Redrawn path diagram of the model in Figure 1.6 including structural coefficients. *Note*: Standardized coefficients are enclosed in parentheses and significant coefficients are indicated by *.

so that

$$TE_{Y_3 X_2} = DE_{Y_3 X_2} + IE_{Y_3 X_2} = 0.103 + 0.114 = 0.217, \qquad (1.45)$$

which corresponds to the entry in Table 1.14.

What, if any, interpretative gain is obtained from estimating and studying the various structural effect components? The focus of the present example is the association between high school rank and highest held degree ($\hat{\rho}_{Y_3 X_2} = 0.189$ from the input matrix in Table 1.13). If standardized coefficients in Table 1.14 are used to estimate the various components of the effect of X_2 (*HSRank*) on Y_3 (*Degree*), one obtains $DE = 0.083$, $IE = 0.092$, and $TE = 0.175$. It can be concluded that—*assuming the particular hypothesized structure of the path model in Figure 1.6 is indeed correct*—92.6% (0.175/0.189) of the association between the two variables is due to some kind of structural effect. The remaining 7.4% are attributable to nonstructural effects that involve the correlation between *HSRank* and *FaEd*. More specifically, 43.9% (0.083/0.189) of the total association between *HSRank* and *Degree* is due to a significant direct structural effect, and 48.7% (0.092/0.189) of the correlation is due to a significant indirect effect through the intervening variables degree aspiration and college selectivity.

As an illustration of the care that must be taken in interpreting structural effect components, compare the relationship between father's educational level and respondent's highest held degree ($\hat{\rho}_{Y_3 X_1} = 0.129$) across the different models in Figures 1.4 and 1.7. In the context of the model in Figure 1.4, it was shown by equations (1.34)–(1.36) that for the two variables *FaEd* and *Degree*, $DE = .045$, $IE = 0.084$, and $TE = 0.129$. For the same two variables in the model depicted in Figure 1.7,

$$DE = 0.044, \tag{1.46}$$

$$IE = 0.062, \tag{1.47}$$

and

$$TE = 0.106. \tag{1.48}$$

The model in Figure 1.4 does *not* include the exogenous variable high school rank, and 100% of the correlation between *FaEd* and *Degree* is "explained" by the total structural effect. The model in Figure 1.7 *does* include *HSRank*, but only about 82% (0.106/0.129) of $\hat{\rho}_{Y_3 X_1}$ is explained by the total effect. Involving a second exogenous variable (*HSRank*) that has a non-zero correlation with *FaEd reduces* that part of the association between *FaEd* and *Degree* which is attributable to the particular hypothesized structure. However, the addition of *HSRank* into the model *increases* the overall amount of explained variability in the endogenous variable *Degree* from 10.8% ($R^2_{\text{sub}} = 0.108$) to 11.4% ($R^2_{\text{full}} = 0.114$). This observed increase in the measure of data-model fit (the coefficient of determination) is statistically significant as can be verified with the *F*-test in equation (1.17), i.e.,

$$F = \frac{(0.114 - 0.108)/(4 - 3)}{(1 - 0.114)/(3094 - 4 - 1)} = 20.919$$

with $df_{\text{num}} = 1$ and $df_{\text{den}} = 3089$.

Finally, compare the relative contribution of the indirect structural effects of father's educational level on respondent's highest held degree to the overall correlation between the two variables across the two models. For the model that does not include the exogenous variable *HSRank* (Figure 1.4), about 65% (0.084/0.129) of the correlation is due to indirect effects through the intervening variables degree aspiration and college selectivity [see equation (1.35)], while from equation (1.47) it can be concluded that, for the model in Figure 1.7, the corresponding percentage is only about 48% (0.062/0.129). These differences in contribution of the total or indirect effects to the overall correlation, however, cannot be interpreted as an indication of the superiority of one over the other model; they merely are reflections of the different hypothesized structures—each of which may be reflecting the "true" relationships equally accurately or inaccurately.

In summary, the estimation of various effect components might help the researcher to understand and interpret the observed associations between variables *within a given path model*. It is important to recognize that all effect estimates are model-dependent; that is, *a different model, derived from a different theoretical foundation and possibly equally justifiable, leads to different effect estimates. No statistical comparison can be used to determine which path model is the one that best represents the true structure among the observed variables* (unless the models are nested; see Chapter 2).

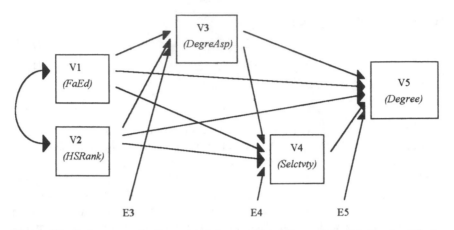

Figure 1.8. Redrawn path diagram of the model in Figure 1.6 with Bentler-Weeks notation.

EQS Example 1.4: The Direct and Indirect Structural Effects in a Path Model of Father's Educational Level and Respondent's High School Rank on Academic Degree. For reference purposes, Figure 1.8 shows the path diagram corresponding to the model in Figure 1.6 in the Bentler-Weeks notational system. Recall that in EQS observed Variables are denoted by the letter V and Error terms associated with endogenous variables are indicated by the letter E.

EQS Input. The EQS program for the model in Figure 1.8 is listed in Table 1.15. Statements that have not been discussed previously are explained in the following:

Line 4: In the required /SPECIFICATIONS section, the statement "Analysis = Covariance;" instructs EQS to base its parameter estimates on an analysis of the variance/covariance matrix of observed variables irrespective of the input matrix. Similar to LISREL, "Analysis = Covariance;" is the default in EQS and is included here for illustrative purposes only.

Lines 21 and 22: /STANDARD DEVIATIONS section (conditional). Since a correlation matrix is used as input data (Matrix = Correlation; on line 4) but a covariance matrix is being analyzed (Analysis = Covariance; on line 4), the standard deviations of observed variables must be given (line 22).

Lines 23 and 24: /PRINT section (conditional). This section of an EQS file allows the user to request indirect and total effect components based on both metric and standardized structural coefficients. This is indicated by the statement "Effects = yes;" on line 24.

Selected EQS Output. The portion of the EQS output that pertains to the direct, indirect, and total effects among variables in the model of Figure 1.8 is

Table 1.15. EQS Input File for the Model in Figure 1.8

1	/TITLE
2	Example 1.4. Path Analysis With Two Exogenous Variables
3	/SPECIFICATIONS
4	Cases = 3094; Variables = 5; Matrix = Correlation; Analysis = Covariance;
5	/LABELS
6	V1 = FaEd; V2 = HSRank; V3 = DegreAsp; V4 = Selctvty; V5 = Degree;
7	/EQUATIONS
8	V3 = *V1 + *V2 + E3;
9	V4 = *V1 + *V2 + *V3 + E4;
10	V5 = *V1 + *V2 + *V3 + *V4 + E5;
11	/VARIANCES
12	V1 = *; V2 = *; E3 = *; E4 = *; E5 = *;
13	/COVARIANCES
14	V2,V1 = *;
15	/MATRIX
16	1
17	.128 1
18	.122 .194 1
19	.300 .372 .214 1
20	.129 .189 .253 .254 1
21	/STANDARD DEVIATIONS
22	1.511 .777 1.014 1.990 .962
23	/PRINT
24	Effects = Yes;
25	/END

shown in Table 1.16. All direct effects are given at the beginning of the output and are the (metric or standardized) structural coefficients in the equations of the dependent variables $DegreAsp$ (V3), $Selctvty$ (V4), and $Degree$ (V5). As before, associated standard errors and t-values are given below each coefficient.

Total and indirect effects (based on both metric and standardized coefficients) are listed in equation form on the EQS output: In an equation associated with a particular dependent variable, a numerical value printed in front of an independent variable indicates either the total or indirect effect of that variable on the particular dependent variable. For example, consider the unstandardized indirect effect of $HSRank$ (V2) on $Degree$ (V5) through all intervening variables [$DegreAsp$ (V3) and $Selctvty$ (V4)], previously calculated as 0.114 in equation (1.43). In the output section entitled DECOMPOSITION OF EFFECTS WITH NONSTANDARDIZED VALUES— PARAMETER INDIRECT EFFECTS, this value is the coefficient associated with V2 in the V5 equation. The corresponding standardized indirect effect of $HSRank$ on $Degree$ (0.092) can be found in the section DECOMPOSITION OF EFFECTS WITH STANDARDIZED VALUES.

Table 1.16. Selected EQS Output from the Analysis of the Model in Figure 1.8

MAXIMUM LIKELIHOOD SOLUTION (NORMAL DISTRIBUTION THEORY)

MEASUREMENT EQUATIONS WITH STANDARD ERRORS AND TEST
STATISTICS

DegreAsp = V3 =   .066*V1 +   .237*V2 +   1.000 E3
                    .012   .023
                    5.582   10.248

Selctvty = V4 =   .241*V3 +   .322*V1 +   .812*V2 +   1.000 E4
                    .032   .021   .042
                    7.529   15.167   19.433

Degree = V5 =   .186*V3 +   .081*V4 +   .028*V1 +   .103*V2 +   1.000 E5
                    .017   .009   .011   .023
                    11.174   8.783   2.480   4.508

DECOMPOSITION OF EFFECTS WITH NONSTANDARDIZED VALUES

PARAMETER TOTAL EFFECTS

DegreAsp = V3 =   .066*V1 +   .237*V2 +   1.000 E3

Selctvty = V4 =   .241*V3 +   .338*V1 +   .869*V2 +   .241 E3 +   1.000 E4

Degree = V5 =   .205*V3 +   .081*V4 +   .068*V1 +   .217*V2 +   .205 E3 +
                    .081 E4 +   1.000 E5

PARAMETER INDIRECT EFFECTS

Selctvty = V4 =   .016*V1 +   .057*V2 +   .241 E3
                    .004   .009   .032
                    4.473   6.052   7.523

Degree = V5 =   .020*V3 +   .040*V1 +   .114*V2 +   .205 E3 +   .081 E4
                    .003   .004   .010   .016   .009
                    5.716   9.128   11.146   12.465   8.783

DECOMPOSITION OF EFFECTS WITH STANDARDIZED VALUES

PARAMETER TOTAL EFFECTS

DegreAsp = V3 =   .099*V1 +   .181*V2 +   .976 E3

Selctvty = V4 =   .123*V3 +   .257*V1 +   .339*V2 +   .120 E3 +   .885 E4

Degree = V5 =   .216*V3 +   .168*V4 +   .107*V1 +   .175*V2 +   .211 E3 +
                    .149 E4 +   .942 E5

PARAMETER INDIRECT EFFECTS

Selctvty = V4 =   .012*V1 +   .022*V2 +   .120 E3

Degree = V5 =   .021*V3 +   .062*V1 +   .092*V2 +   .211 E3 +   .149 E4

STANDARDIZED SOLUTION:

DegreAsp = V3 =   .099*V1 +   .181*V2 +   .976 E3

Selctvty = V4 =   .123*V3 +   .244*V1 +   .317*V2 +   .885 E4

Degree = V5 =   .196*V3 +   .168*V4 +   .044*V1 +   .083*V2 +   .942 E5

For another illustration, consider the total unstandardized structural effect of the exogenous variable *FaEd* (V1) on the endogenous variable *Selctvty* (V4). Its value, 0.338, is the coefficient associated with V1 in the V4 equation listed under DECOMPOSITION OF EFFECTS WITH NONSTANDARDIZED VALUES—PARAMETER TOTAL EFFECTS. This estimate also can be computed from the direct and indirect effect components of V1 on V4: The direct effect, $DE = 0.322$, is the structural coefficient associated with V1 in the V4 equation given at the beginning of the EQS output, while the indirect effect, $IE = 0.016$, is given by the coefficient of V1 in the V4 equation in the output section entitled PARAMETER INDIRECT EFFECTS. Thus, $0.322 + 0.016 = 0.338$ is the total unstandardized effect of *FaEd* on *Selctvty*.

Finally, standard errors and *t*-values for all indirect effects are listed underneath each unstandardized indirect effect; note, however, that standard errors and *t*-values for total effect are not given by EQS. A comparison of results in Tables 1.14 and 1.16 shows that LISREL and EQS estimates of effect components are equal within rounding error.

Identification

A serious problem during the estimation of coefficients in structural equation models is that of model underidentification. The *identification* of a path model refers to the question of whether or not the researcher has sufficient variance and covariance information from the observed variables to estimate the unknown coefficients in the matrices **B**, **Γ**, **Φ**, and **Ψ**. A model is said to be *identified* if all unknown parameters are identified; an individual parameter is said to be identified if it can be expressed as a function of the variances and covariances of the observed variables in the model. More specifically, a structural equation model can have one of three identification statuses:

1. *Just-identified*. A system of equations of the model-implied parameters and the variances and covariances of observed variables can be uniquely solved for the unknown model parameters. This leads to a unique set of parameter estimates once sample variances and covariances of observed variables are available.
2. *Overidentified*. The system of equations can be solved for the model-implied parameters (in terms of the variances and covariances of observed variables), but for at least one such parameter there is no *unique* solution. Rather, based on the variances and covariances among observed variables, there exist at least two solutions for the same parameter. This leads to an implicit, necessary (and testable) equality constraint among certain variance/covariance parameters before structural coefficients can be estimated from sample data.

3. *Underidentified.* The system of equations cannot be solved for all model parameters due to an insufficient number of variances and covariances of observed variables. Now, some parameters cannot be estimated on the sole basis of sample data from the observed variables.

The task of establishing the identification status of a particular path or general structural equation model is not an easy one, since the researcher needs to investigate whether or not *each* parameter can be written as a function of the variances and covariances of observed variables. Fortunately, SEM programs such as LISREL and EQS include algorithms that usually warn the user if a certain coefficient might not be identified. Bollen (1989) and others discussed several ways for checking identification that are beyond the scope and purpose of this book. Nevertheless, to sensitize the reader to the potential problem, a short demonstration of some of the steps required for establishing model identification follows.

One necessary *but not sufficient* condition for the identification of any type of structural equation model is that the number of nonredundant variances and covariances among observed variables, $c = (NX + NY) \times (NX + NY + 1)/2$, is greater than or equal to the number of parameters to be estimated in the model, p. That is, if $c < p$, a structural equation model is underidentified but $c \geq p$ does not necessarily imply that the model is identified. Recursive path analysis models with a diagonal $\boldsymbol{\Psi}$ matrix (no non-zero covariances between error terms of endogenous variables) are never underidentified and are just-identified if and only if $c = p$. For example, the model in Figure 1.2 is just-identified since there are ten nonredundant variances and covariances among the four variables ($c = [1 + 3][1 + 3 + 1]/2$) and ten coefficients in the four basic matrices, $\mathbf{B}$, $\boldsymbol{\Gamma}$, $\boldsymbol{\Phi}$, and $\boldsymbol{\Psi}$, that need to be estimated ($\beta_{21}, \beta_{31}, \beta_{32}, \gamma_{11}, \gamma_{21}, \gamma_{31}, \sigma_{X_1}^2, \sigma_{\zeta_1}^2, \sigma_{\zeta_2}^2, \sigma_{\zeta_3}^2$). Similarly, the model in Figure 1.6 is just-identified since $c = p = 15$.

A Just-Identified Model

Suppose for the purpose of this illustration that one of the research tasks in the study of the relationship between father's educational level (*FaEd*) and respondent's highest held academic degree (*Degree*) is to estimate the direct and indirect effects through the single intervening variable degree aspiration (*DegreAsp*) without regard to the other variables considered in previous examples. The path diagram for this reduced model is shown in Figure 1.9.

This model can be represented by the four basic matrices

$$\mathbf{B} = \begin{bmatrix} 0 & 0 \\ \beta_{21} & 0 \end{bmatrix}, \quad \boldsymbol{\Gamma} = \begin{bmatrix} \gamma_{11} \\ \gamma_{21} \end{bmatrix}, \quad \boldsymbol{\Phi} = [\sigma_{X_1}^2], \quad \text{and} \quad \boldsymbol{\Psi} = \begin{bmatrix} \sigma_{\zeta_1}^2 & 0 \\ 0 & \sigma_{\zeta_2}^2 \end{bmatrix}.$$

Since the number of nonredundant variances and covariances of observed variables is $c = (1 + 2)(1 + 2 + 1)/2 = 6$ and the number of model implied

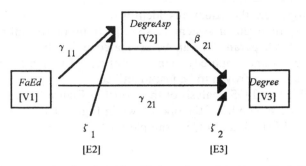

Figure 1.9. An example of a just-identified path model. *Note*: Variable names in brackets correspond to the Bentler-Weeks notation used in EQS.

parameters is $p = 6$, the model must be just-identified. That is, a system of six equations—one corresponding to each variance and covariance of the observed variables—exists that involves the six model-implied parameters $(\beta_{21}, \gamma_{11}, \gamma_{21}, \sigma_{X_1}^2, \sigma_{\zeta_1}^2, \text{ and } \sigma_{\zeta_2}^2)$ and the variances/covariances of the considered variables:

1. To establish the identification of the three structural parameters, β_{21}, γ_{11}, and γ_{21}, one can use the First Law of Path Analysis in equation (1.27) to decompose each nonredundant *covariance* among variables in the model. It is left to the reader (Exercise 1.9) to verify that this application of the first law yields the three equations

$$\sigma_{Y_1 X_1} = \gamma_{11} \sigma_{X_1}^2, \tag{1.49}$$

$$\sigma_{Y_2 Y_1} = \beta_{21} \sigma_{Y_1}^2 + \gamma_{21} \sigma_{Y_1 X_1}, \tag{1.50}$$

$$\sigma_{Y_2 X_1} = \beta_{21} \sigma_{Y_1 X_1} + \gamma_{21} \sigma_{X_1}^2. \tag{1.51}$$

Equations (1.49)–(1.51) represent a system of three equations in three unknowns which can be solved uniquely for the three structural coefficients. That is, if sample estimates of the variances and covariances are available, the unknown structural parameters can be estimated uniquely.

2. To verify the identification of the remaining model-implied parameters $(\sigma_{X_1}^2, \sigma_{\zeta_1}^2, \text{ and } \sigma_{\zeta_2}^2)$, the First Law of Path Analysis [equation (1.27)] is applied to the *variances* of the observed variables. That is,

$$\sigma_{X_1}^2 = \sigma_{X_1}^2, \tag{1.52}$$

$$\sigma_{Y_1}^2 = \gamma_{11} \sigma_{Y_1 X_1} + \sigma_{Y_1 \zeta_1} = \gamma_{11} \sigma_{Y_1 X_1} + \sigma_{\zeta_1}^2, \tag{1.53}$$

$$\sigma_{Y_2}^2 = \beta_{21} \sigma_{Y_2 Y_1} + \gamma_{21} \sigma_{Y_2 X_1} + \sigma_{Y_2 \zeta_2} = \beta_{21} \sigma_{Y_2 Y_1} + \gamma_{21} \sigma_{Y_2 X_1} + \sigma_{\zeta_2}^2. \tag{1.54}$$

Again, there are three equations in only three unknown model parameters $(\sigma_{X_1}^2, \sigma_{\zeta_1}^2, \text{ and } \sigma_{\zeta_2}^2)$ since β_{21}, β_{11}, and γ_{21} have been shown to be identified in step 1. Thus, the elements in Φ and Ψ also are just-identified.

Taken together, the equations identified in steps 1 and 2 [equations (1.49)–(1.54)] represent a system of six equations in six unknown model parameters. Each parameter can be estimated uniquely from sample estimates of the variances and covariances of observed variables; the model in Figure 1.9 has been proven to be just-identified.

A LISREL and EQS analysis of the model in Figure 1.9 is not provided. The model is a submodel of the one shown in Figure 1.1 that was discussed previously in LISREL and EQS Example 1.3.

An Overidentified Model

Now suppose that there is enough theoretical evidence to hypothesize that the direct structural effect of *FaEd* on *Degree* in Figure 1.9 is 0. A path diagram of this situation is shown in Figure 1.10.

The four basic matrices for this model are

$$\mathbf{B} = \begin{bmatrix} 0 & 0 \\ \beta_{21} & 0 \end{bmatrix}, \quad \mathbf{\Gamma} = \begin{bmatrix} \gamma_{11} \\ 0 \end{bmatrix}, \quad \mathbf{\Phi} = [\sigma_{X_1}^2], \quad \text{and} \quad \mathbf{\Psi} = \begin{bmatrix} \sigma_{\zeta_1}^2 & 0 \\ 0 & \sigma_{\zeta_2}^2 \end{bmatrix}.$$

There still are $c = 6$ nonredundant variances and covariances among observed variables, but only five parameters need to be estimated ($\beta_{21}, \gamma_{11}, \sigma_{X_1}^2, \sigma_{\zeta_1}^2$, and $\sigma_{\zeta_2}^2$). The fact that $c > p$ implies that the model in Figure 1.10 might be identified. To see why this model actually is overidentified, apply the First Law of Path Analysis [equation (1.27)] to all nonredundant variances and covariances of observed variables. The reader is asked in Exercise 1.9 to verify that this application of the first law yields the following six equations:

$$\sigma_{Y_1 X_1} = \gamma_{11} \sigma_{X_1}^2, \tag{1.55}$$

$$\sigma_{Y_2 Y_1} = \beta_{21} \sigma_{Y_1}^2, \tag{1.56}$$

$$\sigma_{Y_2 X_1} = \beta_{21} \sigma_{Y_1 X_1}, \tag{1.57}$$

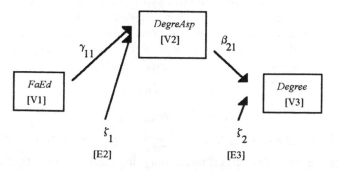

Figure 1.10. An example of an overidentified path model. *Note*: Variable names in brackets correspond to the Bentler-Weeks notation used in EQS.

$$\sigma_{X_1}^2 = \sigma_{X_1}^2, \tag{1.58}$$

$$\sigma_{Y_1}^2 = \gamma_{11}\sigma_{Y_1 X_1} + \sigma_{Y_1 \zeta_1} = \gamma_{11}\sigma_{Y_1 X_1} + \sigma_{\zeta_1}^2, \tag{1.59}$$

$$\sigma_{Y_2}^2 = \beta_{21}\sigma_{Y_2 X_1} + \sigma_{Y_2 \zeta_2} = \beta_{21}\sigma_{Y_2 Y_1} + \sigma_{\zeta_2}^2. \tag{1.60}$$

The system represented by equations (1.58)–(1.60) indicates that, if the structural coefficients (β_{21} and γ_{11}) are identified, the unknown elements in $\mathbf{\Phi}$ and $\mathbf{\Psi}$ are just-identified. Furthermore, equation (1.55) implies that γ_{11} is just-identified. However, the remaining two equations show that the structural coefficient β_{21} is overidentified: Equation (1.56) implies that $\beta_{21} = \sigma_{Y_2 Y_1}/\sigma_{Y_1}^2$, while equation (1.57) shows that $\beta_{21} = \sigma_{Y_2 X_1}/\sigma_{Y_1 X_1}$. Thus, it must be tested whether or not the following equality constraint holds:

$$\sigma_{Y_2 Y_1}/\sigma_{Y_1}^2 = \sigma_{Y_2 X_1}/\sigma_{Y_1 X_1} \tag{1.61}$$

before a sample estimate of β_{21} can trusted.

LISREL EXAMPLE 1.5: AN OVERIDENTIFIED MODEL OF FATHER'S EDUCATIONAL LEVEL ON ACADEMIC DEGREE. Overidentifying restrictions such as the one in equation (1.61) can be statistically tested. An appropriate chi-square test with $df = c - p$ (the number of nonredundant variances and covariances among observed variables minus the number of estimated model-implied parameters) is readily available in LISREL (also see Chapter 2). The appropriate LISREL input program for the model in Figure 1.10 is presented in Table 1.17.

All LISREL syntax, with the exception of the statement on line 10, was introduced in previous examples and requires no further explanation. As a default, LISREL assumes that *all* elements in the structural coefficient matrix $\mathbf{\Gamma}$ are to be estimated. For the example in Figure 1.10, both elements in $\mathbf{\Gamma}$ (γ_{11} and γ_{21}) are assumed to be "free" rather than "fixed" to some constant, e.g., 0. Since the direct effect of *FaEd* on *Degree* is hypothesized to be absent,

Table 1.17. LISREL Input File for the Model in Figure 1.10

1	Example 1.5. An Overidentified Model
2	DA NI = 3 NO = 3094
3	LA
4	DegreAsp Degree FaEd
5	CM SY
6	1.028
7	.247 .925
8	.187 .188 2.283
9	MO NY = 2 NX = 1 BE = SD GA = FU PH = SY PS = DI
10	FI GA(2,1)
11	OU ND = 3

Table 1.18. Selected LISREL Output from the Analysis of
the Model in Figure 1.10

LISREL ESTIMATES (MAXIMUM LIKELIHOOD)

BETA

	DegreAsp	Degree
DegreAsp	—	—
Degree	0.240 (0.017) 14.560	—

GAMMA

	FaEd
DegreAsp	0.082 (0.012) 6.839
Degree	—

PHI

FaEd
2.283

PSI

DegreAsp	Degree
1.013 (0.026) 39.319	0.866 (0.022) 39.319

GOODNESS OF FIT STATISTICS

CHI-SQUARE WITH 1 DEGREES OF FREEDOM = 32.691 (P = 0.0)

---

element (2,1) of the **GA** matrix must be set, or FIxed, to 0. Program line 10 in
Table 1.17—FI GA(2,1)—accomplishes this task.

A portion of the LISREL output is shown in Table 1.18. Even though
parameter estimates are given (e.g., $\hat{\beta}_{21} = 0.240$ and $\hat{\gamma}_{11} = 0.082$), the reported chi-square statistic of 32.691 with $df = 1$ ($p < 0.001$) indicates that the
overidentifying restriction in equation (1.61) might not be true in the population. Equivalently, the chi-square test shows that γ_{21} (the direct effect of $FaEd$
on *Degree*) probably is non-zero, contradicting the hypothesized structure in
Figure 1.10. As is discussed in Chapter 2, the significant chi-square also may
be interpreted as evidence that the data do not fit the presently hypothesized
model and hence the overall model in Figure 1.10 may be rejected.

EQS Example 1.5: An Overidentified Model of Father's Educational
Level on Academic Degree. Table 1.19 contains the EQS program for the
overidentified model in Figure 1.10. The /EQUATIONS section contains the

Table 1.19. EQS Input File for the
Model in Figure 1.10

/TITLE
Example 1.5. An Overidentified Model
/SPECIFICATIONS
Cases = 3094; Variables = 3;
/LABELS
V1 = FaEd; V2 = DegreAsp; V3 = Degree;
/EQUATIONS
V2 = *V1 + E2;
V3 = *V2 + E3;
/VARIANCES
V1 = *; E2 = *; E3 = *;
/MATRIX
2.283
.187 1.028
.188 .247 .925
/END

Table 1.20. Selected EQS Output from the Analysis of the Model in
Figure 1.10

GOODNESS OF FIT SUMMARY

CHI-SQUARE = 32.691 BASED ON 1 DEGREES OF FREEDOM
PROBABILITY VALUE FOR THE CHI-SQUARE STATISTIC IS LESS THAN
  0.001

MAXIMUM LIKELIHOOD SOLUTION (NORMAL DISTRIBUTION
  THEORY)

MEASUREMENT EQUATIONS WITH STANDARD ERRORS AND TEST
  STATISTICS

DegreAsp = V2 =	.082*V1 + .012 6.840	1.000 E2
Degree = V3 =	.240*V2 + .017 14.562	1.000 E3

two structural equations implicit in the path diagram (specifically note the
absence of V1 in the V3 equation). If the input file is specified correctly, EQS
makes it easy to determine the number of parameters to be estimated, p,
which might be helpful in determining the identification status of a model:
Simply count the asterisks (*) in an input file; here, there are five asterisks,
thus $p = 5$ as mentioned before.

A portion of the EQS output for this analysis is shown in Table 1.20.

As before, the chi-square statistic of 32.691 with one degree of freedom ($p < 0.001$) indicates that the overidentifying restriction in equation (1.61) probably does not hold in the population and that the direct effect of *FaEd* (V1) on *Degree* (V3) is not equal to 0. Both estimated structural coefficients ($\hat{\gamma}_{11} = 0.082$ and $\hat{\beta}_{21} = 0.240$) are significantly different from 0 ($t = 6.840$ and $t = 14.562$, respectively) and equal to the ones from the LISREL analysis in Table 1.18.

An Underidentified Model

As a final illustration of the identification issue, change the model in Figure 1.9 to the one shown in Figure 1.11. This time, the error terms ζ_2 and ζ_3 are hypothesized to have a non-zero covariance. Situations like this can arise easily when the researcher has evidence that there might be influences from *outside* the model that are common to two or more endogenous variables, or that different outside influences are correlated. For the model depicted in Figure 1.11, for instance, it could have been hypothesized that mother's educational level and/or high school rank (both variables outside the model shown in Figure 1.9) influence both endogenous variables, respondent's degree aspiration and highest held academic degree. Note that models including correlated error terms should not be analyzed using the ordinary least squares criterion since this violates the OLS assumptions. In Chapter 3, alternative estimation procedures are introduced that allow the researcher to hypothesize non-zero covariances among error terms of endogenous variables in just-identified and overidentified structural equation models.

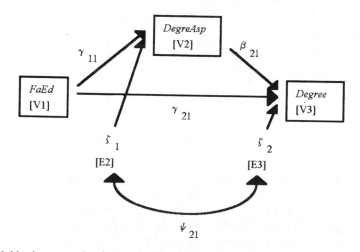

Figure 1.11. An example of an underidentified path model. *Note*: Variable names in brackets correspond to the Bentler-Weeks notation used in EQS.

The model in Figure 1.11 can be represented by the matrices

$$\mathbf{B} = \begin{bmatrix} 0 & 0 \\ \beta_{21} & 0 \end{bmatrix}, \quad \mathbf{\Gamma} = \begin{bmatrix} \gamma_{11} \\ \gamma_{21} \end{bmatrix}, \quad \mathbf{\Phi} = [\sigma_{X_1}^2], \quad \text{and} \quad \mathbf{\Psi} = \begin{bmatrix} \sigma_{\zeta_1}^2 & \sigma_{\zeta_1 \zeta_2} \\ \sigma_{\zeta_2 \zeta_1} & \sigma_{\zeta_2}^2 \end{bmatrix}.$$

As before, the number of nonredundant variances and covariances among variables remains at six, but now there are seven parameters to be estimated (remember that $\sigma_{\zeta_2 \zeta_1} = \sigma_{\zeta_1 \zeta_2}$ in the $\mathbf{\Psi}$ matrix). Thus, the model as a whole cannot be identified. To discover which model-implied parameter(s) could be underidentified, again apply the first law [equation (1.27)] to the six variances and covariances of the observed variables. The resulting system of six equations is as follows (see Exercise 1.9):

$$\sigma_{Y_1 X_1} = \gamma_{11} \sigma_{X_1}^2, \tag{1.62}$$

$$\sigma_{Y_2 Y_1} = \beta_{21} \sigma_{Y_1}^2 + \gamma_{21} \sigma_{Y_1 X_1} + \sigma_{Y_1 \zeta_2} = \beta_{21} \sigma_{Y_1}^2 + \gamma_{21} \sigma_{Y_1 X_1} + \sigma_{\zeta_2 \zeta_1}, \tag{1.63}$$

$$\sigma_{Y_2 X_1} = \beta_{21} \sigma_{Y_1 X_1} + \gamma_{21} \sigma_{X_1}^2, \tag{1.64}$$

$$\sigma_{X_1}^2 = \sigma_{X_1}^2, \tag{1.65}$$

$$\sigma_{Y_1}^2 = \gamma_{11} \sigma_{Y_1 X_1} + \sigma_{Y_1 \zeta_1} = \gamma_{11} \sigma_{Y_1 X_1} + \sigma_{\zeta_1}^2, \tag{1.66}$$

$$\sigma_{Y_2}^2 = \beta_{21} \sigma_{Y_2 Y_1} + \gamma_{21} \sigma_{Y_2 X_1} + \sigma_{Y_2 \zeta_2}$$

$$= \beta_{21} \sigma_{Y_2 Y_1} + \gamma_{21} \sigma_{Y_2 X_1} + (\beta_{21} \sigma_{Y_1 \zeta_2} + \sigma_{\zeta_2}^2) \tag{1.67}$$

$$= \beta_{21} \sigma_{Y_2 Y_1} + \gamma_{21} \sigma_{Y_2 X_1} + (\beta_{21} \sigma_{\zeta_2 \zeta_1} + \sigma_{\zeta_2}^2).$$

Equations (1.62)–(1.64) show that, if $\sigma_{\zeta_2 \zeta_1}$ is identified, the three structural coefficients in $\mathbf{B}$ and $\mathbf{\Gamma}$ (β_{21}, γ_{11}, and γ_{21}) are just-identified since then there are three equations in exactly three unknowns. However, equations (1.65)–(1.67) imply that—even if the structural parameters are identified—not all of the four unknown elements in $\mathbf{\Phi}$ and $\mathbf{\Psi}$ ($\sigma_{X_1}^2$, $\sigma_{\zeta_1}^2$, $\sigma_{\zeta_2}^2$, and $\sigma_{\zeta_2 \zeta_1}$) can be identified (there are only three equations for four unknowns). At least one element in $\mathbf{\Phi}$ or $\mathbf{\Psi}$ must be assumed to be known and given some numerical value—for example, $\psi_{21} = 0$ as in Figure 1.9—before all unknown model-implied parameters can be estimated.

LISREL EXAMPLE 1.6: AN UNDERIDENTIFIED MODEL OF FATHER'S EDUCATIONAL LEVEL ON ACADEMIC DEGREE. Attempting to estimate the parameters of the model in Figure 1.11 with LISREL (see Exercise 1.11) leads to an error message such as the one in Table 1.21: A chi-square with *negative* degrees of freedom should convince any user to reexamine and probably modify the hypothesized model.

Table 1.21. Selected LISREL Output from the Analysis of the Model in Figure 1.11

F_A_T_A_L E_R_R_O_R: Degrees of freedom is negative.

Table 1.22. Selected EQS Output from the Analysis of the Model in Figure 1.11

MAXIMUM LIKELIHOOD SOLUTION (NORMAL DISTRIBUTION THEORY)

PARAMETER	CONDITION CODE
E3,E2	LINEARLY DEPENDENT ON OTHER PARAMETERS
WARNING	TEST RESULTS MAY NOT BE APPROPRIATE DUE TO CONDITION CODE

CHI-SQUARE =　　　　　　0.000 BASED ON −1 DEGREES OF FREEDOM
NONPOSITIVE DEGREES OF FREEDOM. PROBABILITY COMPUTATIONS ARE UNDEFINED.

MEASUREMENT EQUATIONS WITH STANDARD ERRORS AND TEST STATISTICS

DegreAsp = V2 =	.082*V1 + .012 6.840	1.000 E2	
Degree = V3 =	−.655*V2 + .028 −23.470	.136*V1 + .015 8.806	1.000 E3

EQS EXAMPLE 1.6: AN UNDERIDENTIFIED MODEL OF FATHER'S EDUCATIONAL LEVEL ON ACADEMIC DEGREE. It is left to the reader to modify the EQS program in Table 1.19 to analyze the model in Figure 1.11 [see Exercise 1.11; a change of the V3 equation in the /EQUATIONS section is required to estimate the path from V1 (*FaEd*) to V3 (*Degree*), and a /COVARIANCES section must to be added reflecting the hypothesized non-zero covariance between E3 and E2]. Portions of the corresponding EQS output are shown in Table 1.22. Even though parameter estimates, standard errors, and *t*-values are calculated, EQS gives sufficient warning that the user should not attempt to interpret these results. For example, the PARAMETER CONDITION CODE indicates that the covariance between E3 and E2 is linearly dependent on other parameters and thus, may not be identified.

Summary

Working through this first chapter, the reader should have acquired an understanding of the basic ideas behind structural equation modeling with observed variables. Only the application of presented methods to data of the *reader's* interest, however, will ensure that (s)he (a) becomes aware of some of the difficulties and pitfalls usually encountered in constructing theoretically sound and substantively meaningful structural equation models; (b) becomes comfortable with the LISREL, SIMPLIS, and/or EQS command languages; and (c) becomes aware of some of the advantages and disadvantages of working with the two programs. Before studying the next chapters, the reader is encouraged to experiment with SEM analyses of other data sets or at least complete some of the exercises provided at the end of the chapter. For an actual research application, remember to base the analysis on a ratio of at least 10 : 1 of sample size to the number of parameters to be estimated, whenever possible.

This chapter introduced classical path analysis using the concepts underlying univariate multiple linear regression. The latter is a method for partially predicting or explaining a dependent variable, Y, from data of n individuals on one or more independent variables, X_j. In general, a univariate linear regression model can be represented by the matrix equation

$$\mathbf{Y} = \boldsymbol{\alpha} + \mathbf{X}\boldsymbol{\Gamma} + \boldsymbol{\zeta},$$

where $\mathbf{Y}$ is a $(N \times 1)$ column vector of Y-scores; $\boldsymbol{\alpha}$ denotes a $(N \times 1)$ column vector of constant intercept terms; $\mathbf{X}$ is a $(N \times NX)$ data matrix of X-scores; $\boldsymbol{\Gamma}$ denotes a $(NX \times 1)$ column vector of regression coefficients; and $\boldsymbol{\zeta}$ is a $(N \times 1)$ column vector of error terms.

The method of ordinary least squares (OLS) minimizes the sum of the squared differences between observed and model-implied Y-scores. Provided certain statistical assumptions are met, OLS sample estimates of the regression coefficients may be expressed in matrix form by

$$\hat{\boldsymbol{\Gamma}} = (\mathbf{X}'\mathbf{X})^{-1}\mathbf{X}'\mathbf{Y},$$

and the estimate of the intercept term can be computed as

$$\hat{\alpha} = \hat{\mu}_Y - \hat{\gamma}_{Y1}\hat{\mu}_1 - \hat{\gamma}_{Y2}\hat{\mu}_2 - \cdots - \hat{\gamma}_{YNX}\hat{\mu}_{NX}.$$

Before regression coefficients may be used in helping to assess the relative importance of the independent variables within an equation, the coefficients must be standardized. Also, the correlation between observed and model-implied (predicted) scores of the dependent variable—the coefficient of determination, R^2—represents a descriptive measure of the amount of variance of the dependent variable that can be attributed to variability in the independent variable(s). For inferential purposes, an adjusted coefficient of determination, R^2_{adj}, should be computed since an unadjusted R^2 is a biased estimator of the proportion of explained variability in the population. A particular R^2 or R^2_{adj} may be used as an indication of the data-model fit, that is, a measure of the consistency between the observed data and the hypothesized regression model (see Chapter 2).

Structural equation modeling, including classical path analysis, may be used to help bridge the gap between empirical and theoretical research; it is a multivariate statistical technique that uses *empirical* evidence to estimate the strengths of a priori hypothesized structural relationships within a particular *theory-derived* model. The technique enables the researcher to estimate direct, indirect, and total structural effects that might enhance the understanding of relationships among variables within a specific context. The problem of model identification deals with the question of whether or not unique estimates of model-implied parameters can be obtained from available variance/covariance information of observed variables; a model can be just-, over-, or underidentified. Underidentification is a serious problem for the estimation of parameters in any type of structural equation model and might not be easy to detect.

For recursive path models, regression coefficients based on the OLS criterion can be used to estimate structural coefficients provided OLS assumptions are met. Today, however, commercially available computer programs such as LISREL and EQS allow for the statistical testing of direct, indirect, and total effects using a variety of other, more general estimation techniques that are discussed in Chapter 3. A general path model (of which univariate multiple regression is a special case) can be represented by the matrix equation

$$\mathbf{Y} = \boldsymbol{\alpha} + \mathbf{BY} + \boldsymbol{\Gamma}\mathbf{X} + \boldsymbol{\zeta},$$

where $\mathbf{Y}$ is a $(NY \times 1)$ column vector of endogenous variables; $\mathbf{X}$ denotes a $(NX \times 1)$ column vector of exogenous variables; $\boldsymbol{\alpha}$ is a $(NY \times 1)$ column vector of intercept terms; $\mathbf{B}$ (**Beta** or **BE**) is a $(NY \times NY)$ matrix of structural coefficients from endogenous to other endogenous variables; $\boldsymbol{\Gamma}$ (**Gamma** or **GA**) is a $(NY \times NX)$ matrix of structural coefficients from exogenous to endogenous variables; and $\boldsymbol{\zeta}$ (**Zeta** or **ZE**) denotes a $(NY \times 1)$ column vector of error terms of endogenous variables. In addition, a $(NX \times NX)$ variance/covariance matrix, $\boldsymbol{\Phi}$ (**Phi** or **PH**), of observed exogenous variables and a $(NY \times NY)$ variance /covariance matrix, $\boldsymbol{\Psi}$ (**Psi** or **PS**), of elements in $\boldsymbol{\zeta}$ can be specified. Taken together, the four matrices $\mathbf{B}$, $\boldsymbol{\Gamma}$, $\boldsymbol{\Phi}$, and $\boldsymbol{\Psi}$ completely determine the structure of a specific path model. In this book, the assumption is made that variables are measured as deviation scores from their respective means so that all intercept terms in $\boldsymbol{\alpha}$ are 0, unless otherwise stated.

The software packages introduced here are only two from an array of powerful SEM programs available to the potential user of structural equation modeling techniques. The LISREL user has a choice between a matrix-based and a nonmatrix-based (SIMPLIS) command language. The former conforms to the notation used throughout the chapters of this book (the Jöreskog-Keesling-Wiley approach to represent structural equation models) while the latter is discussed in Appendix A. The EQS package, on the other hand, is based on the Bentler-Weeks model and also requires no explicit matrix specifications for the analysis of a structural equation model.

The technical sophistication of both SEM programs makes the choice between using LISREL or EQS one of personal preference, especially for the beginning user of SEM methods. The SIMPLIS and EQS command languages both offer a very user-friendly environment that lets the researcher conduct analyses of simple as well as complex structural equation models with ease and without having to understand fully the matrix representation of such models. On the other hand, the precondition of conceptualizing the matrix representation of a model *before* LISREL's matrix-based code can be written forces the user not only to understand some of the underlying statistical theory but also to carefully think about and research the proposed structure before analysis. This might discourage "casual" modelers from too quickly estimating a variety of alternative models in an attempt to determine the model that best fits the data or that produces the desired significance results (for further discussion, see Chapter 2). Whichever software package the reader will chose to use in future applications, at least an elementary understanding of the representation approach (Jöreskog-Keesling-Wiley or Bentler-Weeks) might be necessary to fully appreciate the power and elegance of SEM methods.

EXERCISES

1.1. Use either LISREL or EQS and the data from the female subsample in Appendix D to estimate and interpret the simple linear regression of
(a) respondent's socioeconomic status [SES; measured by highest held academic degree (*Degree*)] on parents' SES [measured by mother's educational level (*MoEd*)];
(b) respondent's SES [now indicated by occupational prestige (*OcPrestg*)] on parents' SES [measured by parents' joint income (*PaJntInc*)].
Compute and interpret the unadjusted and adjusted coefficients of determination for the regression in parts (a) and (b). Apply appropriate F-tests to decide whether or not these coefficients are significantly different from 0.

1.2. Use LISREL or EQS and the data from the female subsample in Appendix D to estimate and interpret the multiple linear regression of
(a) respondent's SES [measured by income 5 years after college graduation (*Income*)] on parents' SES [indicated by mother's educational level (*MoEd*)] and respondent's high school rank (*HSRank*);
(b) respondent's SES (again indicated by *Income*) on parents' SES (measured by *MoEd*), high school rank (measured by *HSRank*), college prestige [measured by college selectivity (*Selctvty*)], and respondent's motivation [indicated by a measure of self confidence (*SelfConf*)].
Use an appropriate F-statistic to compare results in parts (a) and (b); that is, test whether or not college prestige and respondent's motivation jointly contributed significantly to the prediction of the respondent's socioeconomic status above and beyond the variables parents' SES and high school rank.

1.3. Use either LISREL, SIMPLIS, or EQS and the data from the female subsample in Appendix D to estimate and interpret the intercept terms and structural coefficients [elements in the α, $\mathbf{B}$, and Γ matrices of equation (1.22)] for the path models shown in
(a) Figure 1.2,
(b) Figure 1.6.

Compare your results from these analyses to the ones for the male subsample presented in Tables 1.10 and 1.14.

1.4. Based on the results obtained in Exercise 1.3(a), numerically verify for the female subsample the decomposition of the covariance between the variables *Degree* and *FaEd* that is shown in equation (1.32) for the male subsample.

1.5. Apply the First Law of Path Analysis [equation (1.27)] to symbolically decompose the covariance of
(a) variables Y_3 and Y_1 in Figure 1.2,
(b) variables Y_3 and X_1 in Figure 1.6,
(c) variables Y_2 and X_2 in Figure 1.6.

1.6. Use the standardized structural coefficients in Figure 1.7 to verify that for the male subsample the structural effects from *FaEd* to *Degree* are $DE = 0.044$, $IE = 0.062$, and $TE = 0.106$.

1.7. For the path model in Figure 1.1 and based on the data from the female subsample in Appendix D, use either LISREL, SIMPLIS, or EQS to estimate the unstandardized indirect structural effect of
(a) *FaEd* on *Degree* through the intervening variable *DegreAsp*,
(b) *FaEd* on *Degree* through the intervening variable *Selctvty*,
(c) *FaEd* on *Degree* through both intervening variables *DegreAsp* and *Selctvty*.
Repeat parts (a) through (c), this time computing the indirect effects based on standardized coefficients, and compare the results to the ones for the male subsample shown as part of equation (1.33).

1.8. Use LISREL, SIMPLIS, or EQS and data from the female subsample (Appendix D) to estimate, test, and interpret all (a) direct, (b) indirect, and (c) total structural effects for the model in Figure 1.6. Compare your results to the ones for the male subsample presented in Table 1.16.

1.9. Apply the First Law of Path Analysis [equation (1.27)] to verify
(a) equations (1.49)–(1.54),
(b) equations (1.55)–(1.60),
(c) equations (1.62)–(1.67).

1.10. For the model depicted in Figure 1.6, use LISREL or EQS and data from the male subsample (Appendix D) to statistically test the overidentifying restriction resulting from
(a) a hypothesized 0 direct effect of variable *HSRank* on *Degree*,
(b) a hypothesized 0 direct effect of *DegreAsp* on *Selctvty*,
(c) a hypothesized 0 direct effect of *FaEd* on *Degree* in addition to both restrictions in parts (a) and (b).

1.11. Use (a) LISREL and (b) EQS to verify that the model in Figure 1.11 is underidentified by producing output files equivalent to the ones partially shown in Tables 1.21 and 1.22.

1.12. Use either LISREL or EQS and data from the male subsample (Appendix D) to verify that the model in Figure 1.6 becomes underidentified if
(a) the error terms associated with the variables *DegreAsp* and *Selctvty* are hypothesized to covary,
(b) all elements in the Ψ matrix are specified to be estimated by the program.

1.13. Conceptualize and theoretically justify a path model different from the ones
presented in this chapter consisting of at least four variables from Appendix D.
Use either LISREL, SIMPLIS, or EQS and data from the male or female
subsample to estimate the parameters in the model and interpret the results.

Recommended Readings

Popular reference books on multiple linear regression specifically written for social
scientists can be consulted to gain a more in-depth understanding of the regression
concepts that underlie classical path analysis. See, for example:

Aiken, L.S., and West, S.G. (1991). *Multiple Regression: Testing and Interpreting
    Interactions.* Newbury Park, CA: Sage.
Cohen, J., and Cohen, P. (1983). *Applied Multiple Regression/Correlation Analysis for
    the Behavioral Sciences* (2nd ed.). Hillsdale, NJ: Lawrence Erlbaum.
Draper, N.R., and Smith, H. (1981). *Applied Regression Analysis* (2nd ed.). New York:
    John Wiley & Sons.
Kleinbaum, D.G., Kupper, L.L., and Muller, K.E. (1988). *Applied Regression Analysis
    and Other Multivariable Methods* (2nd ed.). Boston: PWS-Kent.
Pedhazur, E.J. (1982). *Multiple Regression in Behavioral Research: Explanation and
    Prediction* (2nd ed.). New York: Holt, Rinehart and Winston.

There are some introductory and classical texts on structural equation modeling
that are helpful for gaining a historical perspective on the development of path
analysis:

Asher, H.B. (1983). *Causal Modeling* (2nd ed.). Newbury Park, CA: Sage.
Berry, W.D. (1984). *Nonrecursive Causal Models.* Newbury Park, CA: Sage.
Blalock, H.M. (1964). *Causal Inferences in Nonexperimental Research.* Chapel Hill:
    The University of North Carolina Press.
Davis, J.A. (1985). *The Logic of Causal Order.* Newbury Park, CA: Sage.
Duncan, O.D. (1975). *Introduction to Structural Equation Models.* New York: Aca-
    demic Press.
Kenny, D.A. (1979). *Correlation and Causality.* New York: John Wiley & Sons.

CHAPTER 2

Confirmatory Factor Analysis

Overview and Key Points

Confirmatory factor analysis (CFA) is based on the premise that observable variables are imperfect indicators of certain underlying, or latent, constructs. For example, variables used in the regression and path analytical models of Chapter 1, such as father's education (*FaEd*), degree aspirations (*DegreAsp*), and highest held academic degree (*Degree*), can be thought of as imperfect indicators of the latent constructs parents' socioeconomic status (*PaSES*), general academic motivation (*AcMotiv*), and one's own socioeconomic status (*SES*), respectively. If more than one observed indicator variable is available to measure a particular latent construct, CFA allows the researcher to cluster these variables in prespecified, theory-driven ways to evaluate to what extent a particular data set "confirms" what is theoretically believed to be its underlying structure. Thus, the CFA approach to multivariate data analysis does not let a particular data set dictate, identify, or discover underlying dimensions [as is the case with other variable reduction techniques such as exploratory factor analysis (EFA) or principal components analysis (PCA)]; rather, it requires the researcher to theorize an underlying structure and assess whether the observed data "fits" this a priori specified model. In doing so, CFA provides a framework for addressing some of the problems associated with traditional ways of assessing a measure's validity and reliability. Specifically, the following six key points are discussed in this chapter:

1. Any CFA model is determined by the patterns of 0 and non-zero elements in three basic matrices: **Lambda X** (denoted by Λ_x or **LX**), a matrix of structural coefficients linking the observed and latent variables; **Phi** (Φ or **PH**), the variance/covariance matrix of latent constructs; and **Theta Delta** (denoted by Θ_δ or **TD**), a variance/covariance matrix of measurement errors associated with the observed variables.

2. When the unrestricted variance/covariance matrix of observed variables is expressed as a function of the three basic matrices Λ_x, Φ, and Θ_δ, it is referred to as the model-implied variance/covariance matrix.

3. Identification of CFA models usually can be ensured by (a) holding the number of parameters to be estimated at or below the number of non-redundant variances and covariances of observed variables, (b) assigning the unobserved latent variables an appropriate unit of measurement, and (c) assuming that observed variables that are the only indicator of a particular latent variable are measured without error.

4. Whether or not the collected data are consistent with an a priori specified CFA model can be evaluated by a variety of fit indices that have important advantages as well as disadvantages.

5. Based on an identification of possible model misspecifications from data-model fit results, it might be justified to modify an initially hypothesized CFA model to improve the overall data-model fit.

6. Results from a CFA can assist in the assessment of the validity and reliability of an instrument by redefining traditional definitions of these psychometric concepts from a CFA perspective.

Concepts presented in this chapter are illustrated first by LISREL and EQS analyses of data on selected variables previously introduced in the regression and path analytical models from Chapter 1. These examples serve mainly to portray basic CFA concepts and are used again in Chapter 3 as part of an illustration of a general structural equation model. In a second set of examples, data on a particular personality measure are analyzed for the purpose of evaluating validity and reliability from a CFA perspective. Specifically, confirmatory factor analyses are conducted on the variables *MoEd*, *FaEd*, *PaJntInc*, *HSRank*, *AcAbilty*, *SelfConf*, *DegreAsp*, *Selctvty*, *Degree*, *OcPrestg*, and *Income* from the male subsample ($n = 3094$) of the data set introduced in Chapter 1 (see Appendix D for summary statistics). In addition, data on the thinking-feeling-acting behavior orientations of $n = 167$ residence hall advisors at a large southeastern land-grant university are used (Mueller, 1987; Mueller *et al.*, 1990) to illustrate an application of CFA for the validity and reliability assessment of a particular personality measure, the *Hutchins Behavior Inventory* (*HBI*; Hutchins, 1992). Descriptive statistics (means, standard deviations, and intercorrelations) for the various *HBI* scales (T_f, F_a, A_t, T_c, F_c, and A_c) are shown in Appendix E.

As in Chapter 1, all illustrations include listings of possible LISREL and EQS input files together with any necessary annotations and selected portions of output files (consult Appendix A for corresponding SIMPLIS files). However, mainly due to space limitations, not all models discussed are analyzed by both LISREL and EQS; instead, the omitted analyses are left as exercises for the reader and are presented at the end of the chapter.

Specification and Identification of a CFA Model

The central rationale for developing CFA was expressed as early as the beginning of the 20th century with Spearman's (1904) discovery of (exploratory) factor analysis (EFA): Observed variables are imperfect measures of certain unobservable constructs or *latent variables*. For example, no one today could argue successfully for having developed *the* single measure of human intelligence; instead, we have a choice among multiple imperfect indicators (e.g., several intelligence tests such as the WISC or the Stanford-Binet) of one underlying construct, here, intelligence. Colleges and universities routinely use scores from the verbal and quantitative portions of the SAT as two indicators of overall scholastic aptitude, and sociologists frequently use observable variables such as educational level and income as measures of socioeconomic status (SES).

The following four sections illustrate how a particular confirmatory factor analysis model is specified by the model-implied patterns of 0 and non-zero elements in three basic matrices. A distinction is made between the unrestricted and model-implied variance/covariance matrix of observed variables, and the question of CFA model identification is raised. Finally, a partial LISREL analysis of data introduced in the previous chapter is conducted to augment the theoretical discussions.

Model Specification

Reconsider the exogenous portion of the path analytical model of Figure 1.6. The variable father's education (*FaEd*) was used as one possible indicator of the underlying construct parents' socioeconomic status (*PaSES*). Perusing the list of observed variables in Appendix D reveals that two alternative variables, mother's education (*MoEd*) and parents' joint income (*PaJntInc*), also could have been used for that purpose. On the other hand, the second exogenous variable in Figure 1.6, high school rank (*HSRank*), seems to be the only available observed variable to indicate the construct academic rank (*AcRank*). In typical CFA applications, hypothesized relationships between observed variables (such as *MoEd*, *FaEd*, *PaJntInc*, and *HSRank*) and associated latent constructs (such as *PaSES* and *AcRank*) are represented graphically in path diagrams, as in Figure 2.1. Here, mother's education, father's education, and parents' joint income are shown as indicators of one and the same underlying construct, namely parents' socioeconomic status, while high school rank serves as the sole indicator of general academic rank.

As in the path analytical models of Chapter 1, straight arrows in the figure denote structural relations; i.e., relationships with a prespecified structure or direction. Observed variables usually are enclosed by squares or rectangles, while latent variables are enclosed by circles or ellipses. Arrows pointing to

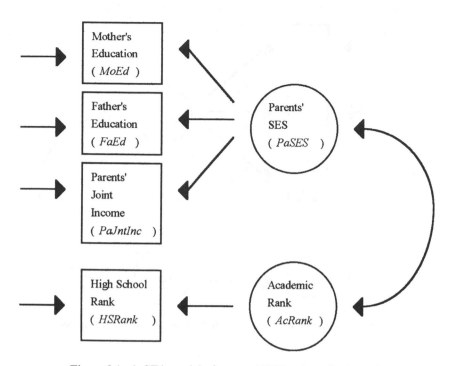

Figure 2.1. A CFA model of parents' SES and academic rank.

variables from "outside" the model (on the left side of Figure 2.1) denote the presence of *measurement errors* in the observed variables. These error terms represent two indistinguishable and unidentified components (one specific to the observed variable, the other pure random measurement error) that, aside from the effects of the underlying latent construct, influence each observed variable.

Before proceeding, a note regarding the direction of the structural relations or "causation" between the observed indicator variables and the latent construct in Figure 2.1 should be made. Bollen (1989) distinguished between *cause indicators* and *effect indicators* of latent variables: The former are defined as observed variables "causing" the underlying latent construct, while the latter should be thought of as being the effects of—that is, are influenced or "caused" by—the latent variables. Thus, in Figure 2.1, *MoEd*, *FaEd*, and *PaJntInc* are specified as effect indicators of *PaSES*, which is contrary to Bollen's (1989, pp. 64–66) opinion that variables such as education, income, and occupational prestige probably should be seen as cause indicators of socioeconomic status. Given the scope of this book, its treatment of the term "cause" (see the Preface), and in accordance with the majority of published applications of CFA, observed indicator variables of latent constructs throughout examples in this book are treated as effect indicators.

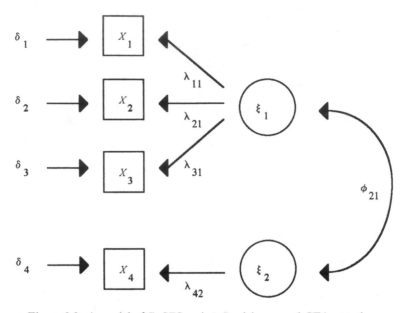

Figure 2.2. A model of *PaSES* and *AcRank* in general CFA notation.

Figure 2.2 depicts the above model in more general CFA notation: Let X_1 denote the observed indicator variable mother's education (*MoEd*), X_2 the variable father's education (*FaEd*), X_3 the observed indicator parents' joint income (*PaJntInc*), and X_4 the variable high school rank (*HSRank*); let ξ_1 (*ksi*) denote the latent underlying construct parents' socioeconomic status (*PaSES*) and ξ_2 be the latent variable academic rank (*AcRank*). Again, note that the fact that the observed variables, $X_j, j = 1, \ldots, NX$ (here, $NX = 4$), are imperfect measures of the NK underlying constructs, $\xi_s, s = 1, \ldots, NK$ (here, $NK = 2$), is made explicit: Measurement errors, δ_j (*delta*), for each observed variable are incorporated in the graphic representation of the CFA model.

If we assume again that the observed variables (X_j) are measured as deviations from their means, the relationships indicated in Figure 2.2 can be expressed in structural equation form as follows:

$$X_1 = \lambda_{11}\xi_1 + \delta_1, \tag{2.1}$$

$$X_2 = \lambda_{21}\xi_1 + \delta_2, \tag{2.2}$$

$$X_3 = \lambda_{31}\xi_1 + \delta_3, \tag{2.3}$$

$$X_4 = \lambda_{42}\xi_2 + \delta_4, \tag{2.4}$$

where each equation expresses an observed variable, X_j (here, mother's education, father's education, parents' joint income, and high school rank), as a function of the latent variables, ξ_s (here, parents' SES and academic rank), and a measurement error term, δ_j. The weights, or structural coefficients, λ_{js} (*lambda*), are factor loadings linearly relating each X_j to the latent variables, ξ_s. In addition, as in Chapter 1, the coefficient ϕ_{21} indicates the hypothesized association between the two exogenous (now latent) variables, ξ_1 and ξ_2. Finally, note that equations (2.1)–(2.4) do not contain intercept terms since the variables were assumed to be measured as deviations from their means (for a discussion on including means in CFA and more general models, consult Bentler, 1993, Chapter 8; or Jöreskog and Sörbom, 1993a, Chapter 10).

As an alternative to representing a CFA model such as the one in Figure 2.1 by listing all its implied structural equations [e.g., equations (2.1)–(2.4)], a model can be represented in matrix form by the equation

$$\mathbf{X} = \boldsymbol{\Lambda}_X \boldsymbol{\xi} + \boldsymbol{\delta}, \tag{2.5}$$

where $\mathbf{X}$ is a $(NX \times 1)$ column vector of observed variables in deviation form; $\boldsymbol{\Lambda}_X$ (**Lambda X** or **LX**) is a $(NX \times NK)$ matrix of structural coefficients; $\boldsymbol{\xi}$ is a $(NX \times 1)$ column vector of latent variables; $\boldsymbol{\delta}$ is a $(NX \times 1)$ column vector of measurement error terms associated with the observed variables; and NX and NK denote the number of observed variables, X_j, and latent constructs, ξ_s, respectively. For the CFA model in Figure 2.2, equation (2.5) becomes

$$\begin{bmatrix} X_1 \\ X_2 \\ X_3 \\ X_4 \end{bmatrix} = \begin{bmatrix} \lambda_{11} & 0 \\ \lambda_{21} & 0 \\ \lambda_{31} & 0 \\ 0 & \lambda_{42} \end{bmatrix} \begin{bmatrix} \xi_1 \\ \xi_2 \end{bmatrix} + \begin{bmatrix} \delta_1 \\ \delta_2 \\ \delta_3 \\ \delta_4 \end{bmatrix}, \tag{2.6}$$

which, as was mentioned above, is equivalent to the set of the four structural equations (2.1)–(2.4).

In addition to the structural coefficient matrix, $\boldsymbol{\Lambda}_X$ (which determines the model-specific relationships between the observed and latent variables), in equation (2.5), two other matrices are needed to completely determine and specify a particular CFA model. Recall that in Figures 2.1 and 2.2 the latent variables $PaSES = \xi_1$ and $AcRank = \xi_2$ are hypothesized to covary (as is indicated by the curved bidirectional arrows in the figures). Thus, similar to the models discussed in Chapter 1, a $(NK \times NK)$ matrix, $\boldsymbol{\Phi}$ (**Phi** or **PH**), of the variances and covariances of latent variables ξ_s can be specified. For the model in Figure 2.2, $\boldsymbol{\Phi}$ can be written as (here and in all subsequent matrix equations, only nonredundant elements are shown)

$$\boldsymbol{\Phi} = \begin{bmatrix} \sigma^2_{\xi_1} \\ \sigma_{\xi_2\xi_1} & \sigma^2_{\xi_2} \end{bmatrix}. \tag{2.7}$$

Finally, consider the measurement errors, δ_j, of the observed variables in Figure 2.2. Since the observed variables *MoEd*, *FaEd*, and *PaJntInc* are assumed to be imperfect measures of the latent construct *PaSES*, the associated measurement error terms, δ_j, should have non-zero variances. On the other hand, it will be argued in the section on identification that the measurement error associated with *HSRank*, δ_4, should be assumed to equal 0 or, at least, be negligible—implying a 0, or near-zero, variance—since the model depicts the variable as the sole indicator of the underlying construct *AcRank*. In any case, the $(NX \times NX)$ variance/covariance matrix, Θ_δ (**Theta Delta** or **TD**), of measurement error terms of the exogenous observed variables is a (4×4) diagonal matrix, listing the variances of the error terms on its diagonal. That is, for the model in Figure 2.2,

$$\Theta_\delta = \begin{bmatrix} \sigma_{\delta_1}^2 & & & \\ 0 & \sigma_{\delta_2}^2 & & \\ 0 & 0 & \sigma_{\delta_3}^2 & \\ 0 & 0 & 0 & \sigma_{\delta_4}^2 \end{bmatrix}. \tag{2.8}$$

In this equation, note the 0 covariances between measurement error terms (the 0 off-diagonal elements in Θ_δ). While measurement errors within an *exploratory* factor analysis must be assumed to be uncorrelated, it is permissible within a CFA framework to specify associations among error terms. The need to do so can arise if, for example, there is sufficient reason to believe that common factors not identified explicitly in the model influence responses to two or more indicator variables. For the time being, however, the absence of any curved bidirectional arrows between the δ_j terms in Figure 2.2 indicates that none of these errors are hypothesized to covary.

In conclusion, the specified patterns of 0 and non-zero elements in the three basic matrices $\Lambda_X = \mathbf{LX}$, $\Phi = \mathbf{PH}$, and $\Theta_\delta = \mathbf{TD}$ completely determine a CFA model. That is, a particular model is specified by (a) identifying the 0 and non-zero structural relations between observed and latent variables in the $(NX \times NK)$ structural coefficient matrix, Λ_X, (b) hypothesizing the (non-structural) associations between the latent factors in the $(NK \times NK)$ variance/covariance matrix, Φ, and (c) stipulating the 0 and non-zero elements in the $(NX \times NX)$ variance/covariance matrix of measurement error terms, Θ_δ.

For the confirmatory factor analyses discussed in this book, certain statistical assumptions must be accepted (the first assumption below can be relaxed for more general models that include mean structures; see Bentler, 1993, Chapter 8; and Jöreskog and Sörbom, 1993a, Chapter 10). As before, when these assumptions are violated, problems could occur with the estimation of model parameters and the testing and interpretation of these estimates. Specifically, for CFA models it is assumed that:

1. The means of the observed and latent constructs are 0, i.e., $E(\mathbf{X}) = E(\xi) = \mathbf{0}$.
2. The relationships between the observed variables, X_j, and the latent constructs, ξ_s, are linear.
3. The measurement error terms in δ in equation (2.5) (a) have a mean of 0 $[E(\delta) = \mathbf{0}]$ and a constant variance across observations; (b) are independent, i.e., uncorrelated across observations; and (c) are uncorrelated with the latent variables $[E(\xi\delta') = E(\delta\xi') = \mathbf{0}]$.

In Figure 2.3, the notation, order, and role of the three basic CFA matrices (Λ_x, Φ, and Θ_δ) are presented schematically. As in Chapter 1, indicator variables, latent constructs, and measurement error terms are minimally connected for ease of reference.

The Unrestricted Versus Model-Implied Variance/Covariance Matrix

In order to facilitate the upcoming discussions of model identification, assessment of data-model fit, and iterative parameter estimation (in Chapter 3), the difference between the unrestricted and the model-implied variance/covariance matrix of observed variables must be explained. First, when no references are made to a particular CFA or more general structural equation model, the population variance/covariance matrix of the observed variables, Σ (**Sigma**), is said to be *unrestricted* since no a priori structure is hypothesized to underlie the variables. Second, recall from equation (2.5) the matrix representation of a CFA model, $X = \Lambda_x \xi + \delta$, and note that Σ can be written as $\Sigma = E(XX')$ since the observed variables are assumed to be measured as deviations from their means $[E(X) = 0]$. Under the additional assumptions that measurement errors in δ have a mean of 0 $[E(\delta) = 0]$ and that these error terms are uncorrelated with the latent variables $[E(\xi\delta') = E(\delta\xi') = 0]$, Σ can be rewritten—by substitution and using the covariance algebra rules reviewed in Appendix B—as

$$\Sigma = E(XX')$$
$$= E[(\Lambda_x\xi + \delta)(\Lambda_x\xi + \delta)']$$
$$= E(\Lambda_x\xi\xi'\Lambda_x') + E(\Lambda_x\xi\delta') + E(\delta\xi'\Lambda_x') + E(\delta\delta')$$
$$= \Lambda_x E(\xi\xi')\Lambda_x' + \Lambda_x E(\xi\delta') + E(\delta\xi')\Lambda_x' + E(\delta\delta')$$
$$= \Lambda_x E(\xi\xi')\Lambda_x' + E(\delta\delta'),$$

where $E(\xi\xi') = \Phi$ and $E(\delta\delta') = \Theta_\delta$ by definition. Thus, the variance/covariance matrix of observed variables can be expressed by

$$\Sigma = \Lambda_x\Phi\Lambda_x' + \Theta_\delta. \tag{2.9}$$

That is, when a particular model is hypothesized to underlie the observed variables, the variance/covariance matrix of observed variables, Σ, can be written as a function of the three basic model-implied matrices, Λ_x, Φ, and Θ_δ. This expression of Σ is referred to as the *model-implied variance/covariance matrix* and, from now on, will be denoted by $\Sigma(\theta)$, where θ is a column vector containing the p model-implied free parameters.

To illustrate the relationship between the unrestricted variance/covariance matrix of observed variables and the model-implied matrix given in equation (2.9), consider the four observed variables X_1 = mother's education, X_2 =

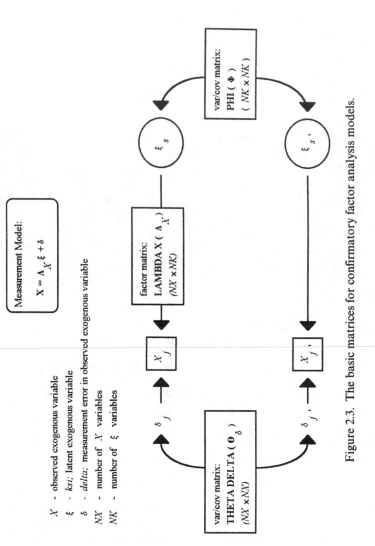

Figure 2.3. The basic matrices for confirmatory factor analysis models.

father's education, $X_3 =$ parents' joint income, and $X_4 =$ high school rank, first *without* any reference to a particular path analytical, CFA, or more general structural equation model. The unrestricted population variance/covariance matrix, Σ, of these four variables is

$$\Sigma = \begin{bmatrix} \sigma^2_{X_1} & & & \\ \sigma_{X_2 X_1} & \sigma^2_{X_2} & & \\ \sigma_{X_3 X_1} & \sigma_{X_3 X_2} & \sigma^2_{X_3} & \\ \sigma_{X_4 X_1} & \sigma_{X_4 X_2} & \sigma_{X_4 X_3} & \sigma^2_{X_4} \end{bmatrix}. \tag{2.10}$$

In contrast, now consider the hypothesized structure (i.e., the restrictions) placed on the four observed variables in the CFA model of Figure 2.2 together with its associated basic matrices Λ_x, Φ, and Θ_δ [in equations (2.6), (2.7), and (2.8)]. By substituting the model-specific forms of Λ_x, Φ, and Θ_δ into equation (2.9), the model-implied variance/covariance matrix, $\Sigma(\theta)$, of the four variables is given by

$$\Sigma(\theta) = \Lambda_X \Phi \Lambda'_X + \Theta_\delta$$

$$= \begin{bmatrix} \lambda_{11} & 0 \\ \lambda_{21} & 0 \\ \lambda_{31} & 0 \\ 0 & \lambda_{42} \end{bmatrix} \begin{bmatrix} \phi_{11} & \\ \phi_{21} & \phi_{22} \end{bmatrix} \begin{bmatrix} \lambda_{11} & 0 \\ \lambda_{21} & 0 \\ \lambda_{31} & 0 \\ 0 & \lambda_{42} \end{bmatrix}' + \begin{bmatrix} \theta_{11} & & & \\ 0 & \theta_{22} & & \\ 0 & 0 & \theta_{33} & \\ 0 & 0 & 0 & \theta_{44} \end{bmatrix}$$

$$= \begin{bmatrix} \lambda_{11} & 0 \\ \lambda_{21} & 0 \\ \lambda_{31} & 0 \\ 0 & \lambda_{42} \end{bmatrix} \begin{bmatrix} \sigma^2_{\xi_1} & \\ \sigma_{\xi_2 \xi_1} & \sigma^2_{\xi_2} \end{bmatrix} \begin{bmatrix} \lambda_{11} & \lambda_{21} & \lambda_{31} & 0 \\ 0 & 0 & 0 & \lambda_{42} \end{bmatrix} + \begin{bmatrix} \sigma^2_{\delta_1} & & & \\ 0 & \sigma^2_{\delta_2} & & \\ 0 & 0 & \sigma^2_{\delta_3} & \\ 0 & 0 & 0 & \sigma^2_{\delta_4} \end{bmatrix}$$

$$= \begin{bmatrix} \lambda^2_{11} \sigma^2_{\xi_1} & & & \\ \lambda_{21} \lambda_{11} \sigma^2_{\xi_1} & \lambda^2_{21} \sigma^2_{\xi_1} & & \\ \lambda_{31} \lambda_{11} \sigma^2_{\xi_1} & \lambda_{31} \lambda_{21} \sigma^2_{\xi_1} & \lambda^2_{31} \sigma^2_{\xi_1} & \\ \lambda_{42} \lambda_{11} \sigma_{\xi_2 \xi_1} & \lambda_{42} \lambda_{21} \sigma_{\xi_2 \xi_1} & \lambda_{42} \lambda_{31} \sigma_{\xi_2 \xi_1} & \lambda^2_{42} \sigma_{\xi_2 \xi_1} \end{bmatrix}$$

$$+ \begin{bmatrix} \sigma^2_{\delta_1} & & & \\ 0 & \sigma^2_{\delta_2} & & \\ 0 & 0 & \sigma^2_{\delta_3} & \\ 0 & 0 & 0 & \sigma^2_{\delta_4} \end{bmatrix}.$$

Thus, the model-implied variance/covariance matrix of *MoEd*, *FaEd*, *PaJntInc*, and *HSRank* in the CFA model of Figure 2.2 is

$$\Sigma(\theta) = \begin{bmatrix} \lambda^2_{11} \sigma^2_{\xi_1} + \sigma^2_{\delta_1} & & & \\ \lambda_{21} \lambda_{11} \sigma^2_{\xi_1} & \lambda^2_{21} \sigma^2_{\xi_1} + \sigma^2_{\delta_2} & & \\ \lambda_{31} \lambda_{11} \sigma^2_{\xi_1} & \lambda_{31} \lambda_{21} \sigma^2_{\xi_1} & \lambda^2_{31} \sigma^2_{\xi_1} + \sigma^2_{\delta_3} & \\ \lambda_{42} \lambda_{11} \sigma_{\xi_2 \xi_1} & \lambda_{42} \lambda_{21} \sigma_{\xi_2 \xi_1} & \lambda_{42} \lambda_{31} \sigma_{\xi_2 \xi_1} & \lambda^2_{42} \sigma_{\xi_2 \xi_1} + \sigma^2_{\delta_4} \end{bmatrix}. \tag{2.11}$$

If the hypothesized CFA model is correct, then $\Sigma = \Sigma(\theta)$, and the right sides of equations (2.10) and (2.11) must be equal. This implies that each variance and covariance of observed variables can be written as a function of the model parameters. For example, from equations (2.10) and (2.11), the variance of parents' joint income (X_3) can be written as

$$\sigma_{X_3}^2 = \lambda_{31}^2 \sigma_{\xi_1}^2 + \sigma_{\delta_3}^2,$$

that is, as a function of (a) the structural coefficient, λ_{31}, linking the variable to its hypothesized underlying construct, parents' socioeconomic status (ξ_1); (b) the variance of the latent variable, $\phi_{11} = \sigma_{\xi_1}^2$; and (c) the variance of the measurement error associated with parents' joint income, $\theta_{33} = \sigma_{\delta_3}^2$. Similarly, the covariance of mother's education (X_1) and high school rank (X_4) is

$$\sigma_{X_4 X_1} = \lambda_{42} \lambda_{11} \sigma_{\xi_2 \xi_1},$$

which expresses this covariance in terms of (a) the structural coefficient, λ_{42}, from the latent variable academic rank (ξ_2) to its indicator variable, high school rank; (b) the structural coefficient, λ_{11}, from the latent construct parents' socioeconomic status to the observed variable, mother's education; and (c) the covariance between the two latent constructs, $\phi_{21} = \sigma_{\xi_2 \xi_1}$.

Identification

Before attempting to estimate the parameters in a CFA model, the identification status of the model needs to be investigated to ensure that sufficient variance/covariance information from the observed variables is available to uniquely estimate the p unknown parameters. In other words, to guarantee the existence and uniqueness of parameter estimates, a sufficient number of restrictions must be placed on the basic matrices Λ_x, Φ, and Θ_δ so that, if θ_1 and θ_2 are two vectors that lead to the same model-implied variance/covariance matrix, i.e., $\Sigma(\theta_1) = \Sigma(\theta_2)$, then the two vectors must be equal. For example, the structural coefficient matrix, Λ_x, for the model of Figure 2.2 [see equation (2.6)] includes several constraints regarding the relationships of observed and latent variables: Each 0 in the matrix represents a restriction that an observed variable does not serve as an indicator of a latent construct [e.g., $\lambda_{22} = 0$ in equation (2.6) indicates that the variable *FaEd* is no indicator of the latent variable *AcRank*]. Furthermore, the variance/covariance matrix of measurement error terms, Θ_δ, is restricted by a priori specifying no nonzero covariances of error terms, that is, all off-diagonal elements in Θ_δ are fixed to 0 [see equation (2.8)].

As was demonstrated in Chapter 1, model identification is a complex issue that is somewhat difficult to prove even for simpler models. Fortunately, however, in most straightforward CFA applications, models can be trusted to be identified when three basic guidelines are followed [see Bollen (1989,

Chapter 7) for an in-depth discussion of the necessary and sufficient conditions for the identification of CFA models]:

1. Recall from Chapter 1 that a necessary—but not sufficient—condition for identification is that the number of parameters to be estimated, p, cannot exceed the number of available nonredundant variances/covariances of observed variables, c, where $c = (NX)(NX + 1)/2$ in a general CFA model.
2. Each latent variable in a model *must* be given a unit of measurement. This can be accomplished in one of two ways: (1) For a given latent variable, ξ_s, specify its scale to equal the scale of one of its indicator variables, X_j; that is, fix one of the structural coefficients, λ_{js}, leading from the latent variable to an associated indicator variable, now termed the *reference variable*, to 1.0; or (2) consider the latent variables as standardized to unit variance, that is, fix the variance, $\phi_{ss} = \sigma_{\xi_s}^2$, of a given latent variable to 1.0.
3. If there is only one observed indicator variable, X, for a given latent variable, ξ, assume the reliability of X to be perfect. That is, assume that the observed variable perfectly measures the latent construct and, thus, set its measurement error variance in the matrix Θ_δ to 0. If the assumption of perfect—or near perfect—reliability is not justifiable, it is best to obtain data from at least one additional indicator variable and include that variable in the model.

Reconsider the CFA model of parents' socioeconomic status and academic rank from Figures 2.1 and 2.2. For that model, the three basic matrices Λ_x, Φ, and Θ_δ are reproduced as follows:

$$\Lambda_X = \begin{bmatrix} \lambda_{11} & 0 \\ \lambda_{21} & 0 \\ \lambda_{31} & 0 \\ 0 & \lambda_{42} \end{bmatrix}, \tag{2.12}$$

$$\Phi = \begin{bmatrix} \sigma_{\xi_1}^2 & \\ \sigma_{\xi_2 \xi_1} & \sigma_{\xi_2}^2 \end{bmatrix}, \tag{2.13}$$

$$\Theta_\delta = \begin{bmatrix} \sigma_{\delta_1}^2 & & & \\ 0 & \sigma_{\delta_2}^2 & & \\ 0 & 0 & \sigma_{\delta_3}^2 & \\ 0 & 0 & 0 & \sigma_{\delta_4}^2 \end{bmatrix}. \tag{2.14}$$

Following the first guideline and counting the non-zero elements in these matrices reveals that there are $p = 11$ parameters in the model that need to be estimated. However, with four observed variables ($X_1 = MoEd$, $X_2 = FaEd$, $X_3 = PaJntInc$, and $X_4 = HSRank$) there are only $c = (4)(4 + 1)/2 = 10$ available variances and covariances [the estimates of the elements in the unrestricted variance/covariance matrix, Σ, see equation (2.10)]. Thus, there is not enough information to estimate all of the unknown model parameters:

The CFA model of parents' socioeconomic status and academic rank—as specified by the patterns in the matrices Λ_x, Φ, and Θ_δ [equations (2.12), (2.13), and (2.14)]—currently is underidentified, i.e., not enough restrictions have been placed on the model to allow parameter estimation. To rectify this problem, follow the second guideline.

Note that the latent variables in Figure 2.2 (*PaSES* and *AcRank*) do not have an assigned unit of measurement, i.e., it is not clear yet in what units the unobserved constructs parents' socioeconomic status and academic rank are being expressed and should be interpreted (inches? seconds? pounds? or, more appropriately, length of formal education? dollars? quartiles?). One way to solve this ambiguity is to designate an indicator variable as the reference variable for each of the latent constructs. For the latent variable parents' SES (ξ_1), for example, one could fix the coefficient λ_{11} in equation (2.12) to 1.0, which would specify the scale of *PaSES* to be equal to the one of mother's education (X_1) that is listed in Appendix D. Alternatively, and equally plausible, either the variable father's education (X_2) or parents' joint income (X_3) could be used as the reference variable. Finally, since the latent construct academic rank (ξ_2) is measured by one observed variable only (*HSRank* = X_4), the latter is the single choice for a reference variable. Accordingly, fix λ_{42} in equation (2.12) to 1.0, which sets the scale of *AcRank* equal to the one listed in Appendix D for the observed variable *HSRank*. Reflecting these changes, the structural coefficient matrix Λ_x from equation (2.12) now becomes

$$\Lambda_x = \begin{bmatrix} 1 & 0 \\ \lambda_{21} & 0 \\ \lambda_{31} & 0 \\ 0 & 1 \end{bmatrix}. \tag{2.15}$$

After these changes, the number of parameters to be estimated in the model is reduced from $p = 11$ to $p = 9$ [see equations (2.13), (2.14), and (2.15)].

A convenient alternative (and the default in LISREL) to identifying reference variables for the purpose of specifying units of measurement for the latent variables is to standardize the constructs to unit variance. If the variances of latent variables (the diagonal elements of the Φ matrix) are fixed to 1.0, the constructs have a unit of measurement equal to their population standard deviations. For the model in Figure 2.2, instead of modifying the structural coefficient matrix, Λ_x, the Φ matrix in equation (2.13) could be changed to

$$\Phi = \begin{bmatrix} 1 & \\ \sigma_{\xi_2 \xi_1} & 1 \end{bmatrix}, \tag{2.16}$$

ensuring that the latent variables parents' socioeconomic status (ξ_1) and academic rank (ξ_2) have a specified unit of measurement (their respective population standard deviations). This modification again reduces the number of parameters to be estimated in the model from $p = 11$ to $p = 9$ [see equations and (2.14), (2.15), and (2.16)].

Finally, consider the measurement error, δ_4, associated with the observed variable high school rank. In the model in Figure 2.1, *HSRank* is the only indicator variable of the latent construct *AcRank*. Here, it seems safe to assume that the variable *HSRank*, has no—or negligible—measurement error; thus, the variance of δ_4 in the Θ_δ matrix can be fixed to 0. Consequently, the pattern of Θ_δ in equation (2.14) is changed to

$$\Theta_\delta = \begin{bmatrix} \sigma_{\delta_1}^2 & & & \\ 0 & \sigma_{\delta_2}^2 & & \\ 0 & 0 & \sigma_{\delta_3}^2 & \\ 0 & 0 & 0 & 0 \end{bmatrix}. \tag{2.17}$$

This final modification leads to a total count of $p = 8$ parameters to be estimated in the three basic matrices Λ_x, Φ, and Θ_δ [see either equations (2.13), (2.15), and (2.17) or equations (2.12), (2.16), and (2.17), depending on how units of measurement are assigned to the latent variables]. Now the number of parameters to be estimated is clearly less than the number of available variances and covariances, $c = 10$; thus, the first guideline is satisfied and the model probably is identified (in fact, it is overidentified).

LISREL EXAMPLE 2.1: A CFA OF PARENT'S SES AND ACADEMIC RANK. The first LISREL illustration of a confirmatory factor analysis is the estimation of parameters in the model of Figure 2.1 based on data from the $n = 3094$ male participants in the 1971 ACE/UCLA Freshman Survey (introduced in Chapter 1). The observed variables used in the analysis are *MoEd*, *FaEd*, *PaJntInc*, and *HSRank*; see Appendix D for specific variable codings and summary statistics. The following two sections present annotated LISREL input and selected output; as before, the corresponding SIMPLIS input and partial output can be found in Appendix A.

LISREL Input. A possible program to estimate the model of Figure 2.1 is shown in Table 2.1. Program lines 1 through 11 contain syntax that was introduced in Chapter 1 and, thus, are not explained here.

Line 12: After specifying the number of observed X variables ($NX = 4$; namely, *MoEd*, *FaEd*, *PaJntInc*, and *HSRank*) and latent *Ksi* constructs ($NK = 2$; namely, *PaSES* and *AcRank*), the MOdel section contains information regarding the overall fixed or free status of the three basic matrices, Λ_x, Φ, and Θ_δ. The statement LX = FU,FI specifies Λ_x to be a FUll matrix containing parameters that all initially are FIxed to 0; PH = SY,FR indicates that the variance/covariance matrix of latent variables, Φ, is a SYmmetric matrix with all parameters FRee to be estimated; while TD = SY,FI specifies the variance/covariance matrix, Θ_δ, of measurement error terms to be a SYmmetric matrix whose elements all are initially FIxed to 0.

Table 2.1. LISREL Input File for the CFA Model in Figure 2.1

1	Example 2.1. CFA of Parents' SES and Academic Rank
2	DA NI = 4 NO = 3094
3	LA
4	MoEd FaEd PaJntInc HSRank
5	KM SY
6	1
7	.610 1
8	.446 .531 1
9	.115 .128 .055 1
10	SD
11	1.229 1.511 2.649 .777
12	MO NX = 4 NK = 2 LX = FU,FI PH = SY,FR TD = SY,FI
13	LK
14	PaSES AcRank
15	VA 1.0 LX(1,1) LX(4,2)
16	FR LX(2,1) LX(3,1)
17	FR TD(1,1) TD(2,2) TD(3,3)
18	OU SC ND = 3

Lines 13 and 14: Together, these lines give descriptive labels to the two latent variables in Figure 2.1, parents' SES and academic rank: The keyword LK (Labels for Ksi variables) on line 13 is followed by the assigned labels of *PaSES* and *AcRank* on line 14.

Line 15: To illustrate the use of reference variables in assigning units of measurement to the latent variables, the structural coefficients from (a) *PaSES* to *MoEd* and (b) *AcRank* to *HSRank* are fixed to 1.0. The statement VA 1.0 LX(1,1) LX(4,2) sets the elements λ_{11} and λ_{42} to a VAlue of 1.0. This assigns the unit of measurement of the observed variable *MoEd* to the latent construct *PaSES* and the unit of *HSRank* to the latent variable *AcRank*. Note that LISREL's default method of assigning units of measurement is to standardize the latent variables to unit variance (see LISREL Example 2.2 later in this chapter).

Line 16: The statement on this program line specifies which elements in the structural coefficient matrix Λ_x should be estimated by the program [recall that this matrix previously was specified to contain all fixed elements (line 12)]. For the model in Figure 2.1, two elements need to be freed: λ_{21} and λ_{31}, the two structural coefficients relating the latent variable *PaSES* to the indicator variables *FaEd* and *PaJntInc*, respectively.

Line 17: Since the variance/covariance matrix of measurement error terms Θ_δ was specified to be a fixed matrix (line 12), the variances of the error terms δ_1, δ_2, and δ_3 associated with *MoEd*, *FaEd*, and *PaJntInc*, respectively, need to be specified as free parameters. The statement FR TD(1,1) TD(2,2) TD(3,3) accomplishes this task. Note again that TD(4,4), the variance of the measurement error in *HSRank*, should stay fixed to 0 since *HSRank* is the only available indicator for the latent construct *AcRank* and, thus, is assumed to be measured without error, i.e., with perfect reliability.

Line 18: Finally, for models involving latent variables, the SC (Standardized Completely) option on the OUtput line requires some comment. The command computes solutions based on standardized observed *and* latent variables. If the user wants to obtain solutions based on standardized latent variables only, i.e., the observed variables remain in their original metric, then the option SS (Standardized Solution) must be specified on the OUtput line.

Selected LISREL Output. Table 2.2 shows parts of the output file corresponding to the LISREL program in Table 2.1. First, the table lists the variance/covariance matrix of the observed variables that LISREL computes from the available product-moment correlations and the standard deviations given in the input file, and then uses them in the estimation of the unknown model parameters. The unstandardized and standardized maximum likelihood estimates of free parameters in the three basic matrices Λ_x, Φ, and Θ_δ are given next. Recall from the discussion of the OUtput line in the previous section (Table 2.1, line 18) that the standardized estimates are based on a standardization of the observed *and* latent variables since the SC (Standardized Completely) option was specified. If, instead, the SS (Standardized Solutions) option would have been used, LISREL would have calculated standardized coefficients based on standardized latent variables but *unstandardized* observed variables.

From the unstandardized estimates in the structural coefficient matrix Λ_x, the structural equations implied in the model [equations (2.1)–(2.4)] now can be estimated by

$$\hat{X}_1 = 1.000\xi_1, \tag{2.18}$$

$$\hat{X}_2 = 1.467\xi_1, \tag{2.19}$$

$$\hat{X}_3 = 1.870\xi_1, \tag{2.20}$$

$$\hat{X}_4 = 1.000\xi_2. \tag{2.21}$$

From the output note also that the estimated coefficients linking the latent variable *PaSES* with its indicator variables *FaEd* and *PaJntInc* ($\hat{\lambda}_{21} = 1.467$ and $\hat{\lambda}_{31} = 1.870$) are statistically significant ($t = 30.355$ and $t = 29.796$, respectively). Similarly, all estimated coefficients in the Φ and Θ_δ matrices have t-values well above 2.0 and, thus, are significantly different from 0 (remember, however, that the significance is due probably to the large sample size of $n = 3094$).

Next, the R^2 values for the four structural equations corresponding to the observed variables in the model [equations (2.18)–(2.21)] are listed. It is possible—and common practice in confirmatory factor analyses to inter-

Table 2.2. Selected LISREL Output from the CFA of the Model in
Figure 2.1

COVARIANCE MATRIX TO BE ANALYZED

	MoEd	FaEd	PaJntInc	HSRank
MoEd	1.510			
FaEd	1.133	2.283		
PaJntInc	1.452	2.125	7.017	
HSRank	0.110	0.150	0.113	0.604

LISREL ESTIMATES (MAXIMUM LIKELIHOOD)

LAMBDA-X

	PaSES	AcRank
MoEd	1.000	—
FaEd	1.467	—
	(0.048)	
	30.355	
PaJntInc	1.870	—
	(0.063)	
	29.796	
HSRank	—	1.000

PHI

	PaSES	AcRank
PaSES	0.774	
	(0.040)	
	19.419	
AcRank	0.098	0.604
	(0.014)	(0.015)
	7.055	39.326

THETA-DELTA

MoEd	FaEd	PaJntInc	HSRank
0.737	0.618	4.312	—
(0.029)	(0.049)	(0.133)	
25.827	12.681	32.361	

SQUARED MULTIPLE CORRELATIONS FOR X-VARIABLES

MoEd	FaEd	PaJntInc	HSRank
0.512	0.729	0.386	1.000

COMPLETELY STANDARDIZED SOLUTION

LAMBDA-X

	PaSES	AcRank
MoEd	0.716	—
FaEd	0.854	—
PaJntInc	0.621	—
HSRank	—	1.000

Table 2.2 (*cont.*)

	PHI			
	PaSES	AcRank		
PaSES	1.000			
AcRank	0.144	1.000		
	THETA-DELTA			
	MoEd	FaEd	PaJntInc	HSRank
	0.488	0.271	0.614	—

pret these coefficients of determination as descriptive reliability estimates for the observed variables. This will be elaborated on later in this chapter. For now, note that, given the specific sample, the indicator variables *MoEd* and *FaEd* have moderate and acceptable reliabilities ($R^2 = 0.512$ and $R^2 = 0.729$, respectively), while the reliability estimate for the variable *PaJntInc* seems somewhat low ($R^2 = 0.386$), probably due to a lack of the young respondents' knowledge regarding their parents' income levels. Further, the variable *HSRank* has perfect reliability ($R^2 = 1.000$) by default since a priori the variance of its measurement error was fixed to 0.

Paralleling the interpretation of standardized structural coefficients in Chapter 1, the estimates of standardized factor loadings in the Λ_x matrix might be useful in determining the relative importance of the observed variables as indicators of their associated latent constructs: Here, the variable *FaEd* seems to be the most reliable and strongest indicator of the latent construct parents' socioeconomic status ($\hat{\lambda}_{11(\text{stand})} = 0.854$), followed by *MoEd* ($\hat{\lambda}_{21(\text{stand})} = 0.716$) and *PaJntInc* ($\hat{\lambda}_{31(\text{stand})} = 0.621$). Also, the estimated correlation between the latent variables *PaSES* and *AcRank* can be found in the standardized Φ matrix, $\hat{\phi}_{21(\text{stand})} = 0.144$. This disattenuated estimate of the association between the two constructs is slightly higher than the attenuated correlations between individual indicator variables (ranging from 0.055 for *PaJntInc* and *HSRank* to 0.115 for *MoEd* and *HSRank*; see the input correlation matrix in Table 2.1) since the effects of measurement errors of the indicator variables of parents' socioeconomic status have been removed.

Prior to demonstrating a confirmatory factor analysis with EQS, it is fitting to introduce the issues of data-model fit and model modification within a confirmatory factor analysis framework. Note that these discussions easily generalize to the structural equation models introduced in Chapter 3. In the next section, the assessment of how well a data set fits a hypothesized CFA model is discussed and illustrated by a continuation of the above LISREL analysis. Then, the issue of model modification is introduced, and an EQS analysis of a CFA model based partially on the endogenous variables from the path analytical model in Figure 1.6 is conducted.

Data-Model Fit

After specifying a particular CFA model and estimating the associated parameters based on a random sample of n individuals (see Chapter 3 for details on various estimation techniques), the question of whether the data fit the hypothesized model should be addressed. Do the collected data provide indications that the hypothesized structure should be rejected (data-model misfit), or is there empirical evidence suggesting that the model, as specified, might be *a* viable representation of the true relationships between observed and latent variables? The issue addressed by this question usually is referred to with the term "model fit." In my opinion, this is an unfortunate choice of words; a more correct term seems to be "data fit" since, in most situations, the question asked should be whether or not the *data* fit an a priori hypothesized model, not whether some *model* fits a particular data set! Also, note the mostly *disconfirmatory*—rather than confirmatory—nature of the data-model fit question: An a priori hypothesized model *can* be rejected based on an observed data-model misfit, but one particular structure *cannot* be confirmed as being *the* right model even if the fit is acceptable. In fact, one data set will fit other alternative structures as well as the one under consideration.

Many overall measures of data-model fit have been suggested in the literature in an attempt to give the user a single criterion by which to judge whether or not a particular data set is consistent with an a priori hypothesized model (for comprehensive reviews consult Marsh *et al.*, 1988, or Tanaka, 1993). Perhaps the most commonly used indices in the applied literature are (a) the chi-square statistic, (b) the goodness-of-fit and adjusted goodness-of-fit indices (Jöreskog and Sörbom, 1981), (c) the normed and nonnormed fit indices (Bentler and Bonett, 1980) and the normed comparative fit index and the nonnormed fit index (Bentler, 1990), and, finally, (d) the parsimonious goodness-of-fit and parsimonious normed fit indices (James *et al.*, 1982; Mulaik *et al.*, 1989). Each such measure has advantages and disadvantages (e.g., they all depend on the multivariate normality assumption) that the researcher should be aware of before rejecting a particular structure or claiming to have constructed a well-fitting model based on any of the available fit indices.

Before presenting an overview of some of the available overall (single-index) measures of fit, it is important to point out that probably the simplest, most straightforward—and sometimes the most valuable—aid in identifying data-model inconsistencies is scrutinizing the individual parameter estimates. A sound understanding of the substantive theory underlying the hypothesized model and elementary statistical knowledge will go a long way in assessing the adequacy of the proposed structure. For example, do the estimated structural coefficients have theory-contradicting signs or magnitudes? Are associated standard errors unusually large? Are there any negative variance estimates? Are the coefficients of determination that are associated with each structural equation (sometimes interpreted as reliability estimates of the observed variables) negative, close to 0, or greater than unity? If any of these basic questions are answered in the affirmative, chances are good that the present data set does not fit at least part of the hypothesized structure or that the model might have been, at least partially, misspecified. Reviewing the partial LISREL output in Table 2.2 reveals (a) no estimated structural coefficients with theory contradicting signs; (b) no large associated standard errors —in fact, they seem small, as indicated by large t-ratios (the parameter estimates divided by their standard errors); and (c) no negative variance estimates in the $\mathbf{\Phi}$ or $\mathbf{\Theta}_\delta$ matrices. However, as was mentioned before, in this sample the coefficient of determination R^2 associated with the variable parents' joint income (*PaJntInc*) is relatively small, indicating that this variable might not have been a very good choice as an indicator of parents' socioeconomic status (*PaSES*). In general, seemingly reasonable parameter estimates alone, however, do not imply a well-fitting data set; overall fit indices that are discussed next can assist in identifying data-model inconsistencies or possible misspecifications that are not detectable by an examination of just the individual parameter estimates.

Overall Measures of Fit

Table 2.3 gives a partial list of the available overall fit indices taken from the LISREL output associated with the program in Table 2.1. In the following sections, each fit index listed in the table first is discussed in general and then interpreted for the specific example analysis.

Table 2.3. Partial Data-Model Fit Output from the LISREL Analysis of the CFA Model in Figure 2.1

GOODNESS OF FIT STATISTICS

CHI-SQUARE WITH 2 DEGREES OF FREEDOM = 7.405 (P = 0.0247)

MINIMUM FIT FUNCTION VALUE = 0.00239

CHI-SQUARE FOR INDEPENDENCE MODEL WITH 6 DEGREES OF FREEDOM = 2627.661

GOODNESS OF FIT INDEX (GFI) = 0.999
ADJUSTED GOODNESS OF FIT INDEX (AGFI) = 0.994
PARSIMONY GOODNESS OF FIT INDEX (PGFI) = 0.200

NORMED FIT INDEX (NFI) = 0.997
NON-NORMED FIT INDEX (NNFI) = 0.994
PARSIMONY NORMED FIT INDEX (PNFI) = 0.332
COMPARATIVE FIT INDEX (CFI) = 0.998

The Chi-Square Statistic

One common and straightforward approach to data-model fit assessment for an overidentified CFA or other structural equation model is to statistically test the overidentifying restriction(s) with the chi-square test that was mentioned in Chapter 1 (see LISREL Example 1.5). Formally, the test evaluates whether or not the unrestricted population variance/covariance matrix of the observed variables Σ is equal to the model-implied variance covariance matrix $\Sigma(\theta)$, i.e., it tests the null hypothesis $H_o: \Sigma = \Sigma(\theta)$. If the specified model is correct, then H_o is true, and each population variance and covariance of observed variables in Σ can be written as a particular additive combination of model-implied parameters that is given in $\Sigma(\theta)$. In that case, the *residual matrix* $\Sigma - \Sigma(\theta)$, is a matrix consisting only of 0's. In practice, however, we do not know the elements of Σ but must estimate them by elements in the sample variance/covariance matrix S. Similarly, we do not know the model parameters in the vector θ and, thus, somehow must estimate these coefficients, too.

In Chapter 3, various estimation methods available in LISREL and EQS are discussed. For now, it suffices to mention that these methods all obtain model parameter estimates—collected in a vector $\hat{\theta}$—by iteratively minimizing a certain function $F[S, \Sigma(\theta)]$ so that S and $\Sigma(\hat{\theta})$ are "close," or, equivalently, that the elements in the residual matrix $S - \Sigma(\hat{\theta})$ are small. Usually, F is called a *fitting function* whose particular form depends on the estimation method used (e.g., F_{ML} is the maximum likelihood fitting function, whereas F_{GLS} denotes the generalized least squares fitting function; see Chapter 3).

Now, under the null hypothesis $H_o: \Sigma = \Sigma(\theta)$ and certain distributional assumptions of the observed data (e.g., multivariate normality if the maximum likelihood method is used), $(n - 1)$ times the minimum value of the fitting function,

$$\chi^2 = (n - 1)F[S, \Sigma(\hat{\theta})] \qquad (2.22)$$

is distributed asymptotically as a χ^2-distribution with $df = (c - p)$, where n denotes sample size; as before, $c = (NX)(NX + 1)/2$ is the number of nonredundant variances and covariances of observed variables, and p is the total number of parameters to be estimated. Thus, *if* $H_o: \Sigma = \Sigma(\theta)$ holds in the population, *if* certain distributional assumptions are met, and *if* the sample size is large enough, the hypothesis of perfect data-model fit can be evaluated with a χ^2-test.

Table 2.3 reports a χ^2-value of 7.405, $p = 0.0247$ for the SES example. The associated degrees of freedom are $df = (10 - 8) = 2$ (with four observed variables, there are $c = 10$ available variances and covariances; after specifying units of measurement for the latent variables *PaSES* and *AcRank*, and fixing the measurement error of the observed variable *HSRank* to 0, there remain $p = 8$ parameters to be estimated; see the previous discussion on identification). Thus, based on a sample of $n = 3094$ respondents and under the assumption of multivariate normality, the null hypothesis of perfect data-model fit must be rejected at the $\alpha = 0.05$ level of significance (but not at $\alpha = 0.01$).

Several shortcomings associated with this formal hypothesis test of data-model fit have been noted in the literature (see Kaplan, 1990, for a review). First, the χ^2-test depends on a number of assumptions (e.g., validity of the tested null hypothesis, multivariate normality of observed variables, and sufficiently large sample size) that, in practical applications, rarely can be argued to be met completely. Second, recall that the adequacy of just-identified models cannot be tested in the above fashion: When no over-identifying restriction is placed on an identified CFA or more general structural equation model, $df = (c - p) = 0$ and $S = \Sigma(\hat{\theta})$ by default so that the null hypothesis holds trivially. Third, more complex models necessarily achieve better data-model fit results than simpler ones, discouraging the researcher to follow the principle of parsimony: If more parameters are free to be estimated, less constraints are placed on the unrestricted variance/covariance matrix, which in turn lowers the value of the fitting function $F[S, \Sigma(\theta)]$ and, hence, the χ^2-statistic.

Fourth, and finally, note that as n increases, the test statistic in equation (2.22) increases, leading to the problem that plausible models might be rejected based on a significant χ^2-statistic even though the discrepancy between S and $\Sigma(\hat{\theta})$ is minimal and unimportant. On the other hand, if the data-model fit question is approached from a strictly disconfirmatory perspective, a statistically powerful test sensitive to even small differences between the unrestricted and model-implied variance/covariance matrices might be exactly what is desired. In the final analysis, rarely—if ever—does a researcher believe that the hypothesized model is *the* "true" model; in most cases, it is known a priori that the specified model is only a simplification and approximation of "reality" and, therefore, in a strict and absolute sense, must be false.

For the SES example, probably not too much emphasis should placed on the significance of the χ^2-statistic ($p = 0.0247$) since, for example, the sample size is rather large ($n = 3094$). Furthermore, the simple (somewhat uninteresting) model in Figure 2.1 is not meant to identify *the* true representation of the latent constructs parents' socioeconomic status and academic rank and their relationship. Rather, it simply is an attempt to identify *a* possible process underlying the four available observed variables (*MoEd*, *FaEd*, *PaJntInc*, and *HSRank*) and is used in Chapter 3 as a way of extending one of the path analytical models of the previous chapter to a full structural equation model with latent variables.

Given the above limitations, Jöreskog and Sörbom (1993b) and others advocated to evaluate a χ^2-value not as a formal test statistic. Instead, they suggested to somewhat informally compare the magnitude of an observed χ^2-value to the mean of its underlying sampling distribution $[E(\chi^2) = df]$; that is, a "small" chi-square value—as compared to its associated degrees of freedom—is indicative of a good fit and a "large" value is an indication of a bad data-model fit (in this sense, the chi-square is a "badness" of fit index as opposed to the "goodness" of fit indices introduced below). In the example, $\chi^2 = 7.405$ and $df = 2$; thus, $\chi^2/df = 3.703$. Whether this ratio is "small" or "large" really depends on the viewpoint of the investigator. No absolute standard has been—and, may be, should be—established, but in practice some interpret ratios as high as 3.00, 4.00, or even 5.00 as still representing a "good" data-model fit (see Bollen, 1989, p. 278; or Marsh et al., 1988, p. 406). Also note that interpretation of the ratio of chi-square over degrees of freedom does not eliminate the issue of sample size dependency of the χ^2-test statistic: Larger samples lead to larger chi-square values which, in turn, lead to larger χ^2/df ratios (since, clearly, $df = c - p$ does not vary as a function of sample size).

The Goodness-of-Fit and Adjusted Goodness-of-Fit Index

Another popular alternative to the hypothesis testing approach to fit assessment of overidentified CFA and more general structural equation models is to determine the amount of observed variance/covariance information that can be accounted for by the hypothesized model. This can be accomplished by using Jöreskog and Sörbom's (1981) Goodness-of-Fit Index (GFI) and Adjusted Goodness-of-Fit Index (AGFI). Both indices should fall between 0 and 1 (although, rarely, negative values are computed) with larger values indicating a better data-model fit. Formally,

$$\text{GFI} = 1 - \frac{F[S, \Sigma(\hat{\theta})]}{F[S, \Sigma(0)]}, \tag{2.23}$$

where the numerator is the minimum value of the fitting function F for the hypothesized model, while the denominator is the minimum value of F when no model is hypothesized (all model parameters in the matrices Λ_x, Φ, and Θ_δ are fixed and, thus, none need to be estimated). That is, the GFI measures "how much *better* the model fits as compared to no model at all" (Jöreskog and Sörbom, 1993b, p. 122).

The AGFI is given by

$$\text{AGFI} = 1 - \left(\frac{c}{df_h}\right) \frac{F[S, \Sigma(\hat{\theta})]}{F[S, \Sigma(0)]}$$

$$= 1 - \frac{c}{df_h}(1 - \text{GFI}), \tag{2.24}$$

where c is the number of nonredundant variances and covariances of observed variables and $df_h = c - p$ are the degrees of freedom for the hypothesized model. It is a shrunken version of the GFI that is adjusted for the model's degrees of freedom in an attempt to "penalize" more complex models (with more parameters to be estimated) with a downward adjustment in the fit index: As the number of free parameters increases, the degrees of freedom decrease, c/df_h increases, and the AGFI decreases; conversely, as the degrees of freedom approach c (corresponding to a decrease in free parameters), the AGFI approaches the GFI in value.

Perhaps the rationale for—and interpretation of—these fit indices can be clarified by summarizing and augmenting Tanaka's (1993) remarks on the explicit parallels between the GFI and AGFI on the one hand and the coefficient of determination R^2 and its adjusted version R^2_{adj} from univariate linear regression on the other hand. First, recall from Chapter 1 that, in an ordinary least squares (OLS) regression, (a) estimation of the predicted scores $\hat{Y}$ involves minimizing the sum of the squared error or residual terms $\Sigma(Y - \hat{Y})^2$, and that (b) the descriptive coefficient of determination R^2 is defined as the correlation between the observed and predicted scores on a dependent variable, $R^2 = \hat{\rho}_{Y\hat{Y}}$. An alternative and equivalent definition of this coefficient is

$$R^2 = 1 - \frac{SS_{res}}{SS_{total}}, \qquad (2.25)$$

where $SS_{total} = (SS_{res} + SS_{reg})$ represents the usual decomposition of total sum of squares $[SS_{total} = \Sigma(Y - \bar{Y})^2]$ into a residual $[SS_{res} = \Sigma(Y - \hat{Y})^2]$ and regression $[SS_{reg} = \Sigma(\hat{Y} - \bar{Y})^2]$ portion (see, for example, Cohen and Cohen, 1983, or Pedhazur, 1982).

Now, when the structural equation modeling GFI in equation (2.23) is compared to the regression R^2 in equation (2.25), the rationale underlying the former measure becomes clear since both indices can be interpreted as sample-specific, descriptive measures of the amount of explained variance information by an a priori hypothesized model. That is, both are of the general form

coefficient

$$= 1 - \frac{\text{minimized residual variance information when a model is specified}}{\text{variance information when no model is specified}}.$$

Second, the adjusted coefficient of determination R^2_{adj} was given in Chapter 1 by the shrinkage formula

$$R^2_{adj} = 1 - \frac{(n - 1)}{(n - NX - 1)}(1 - R^2). \qquad (2.26)$$

It adjusts the value of the R^2 downward to eliminate statistical bias; that is, whereas R^2 should be used as a sample-specific descriptive measure, its adjusted version can be used for inferential purposes. Let the denominator in equation (2.26), $df_h = (n - NX - 1)$, denote the degrees of freedom associated with a hypothesized regression model that incorporates NX predictor variables with a significant effect on the dependent variable. Similarly, consider the numerator, $df_n = (n - 1)$, as the total degrees of freedom associated with not having specified a model at all. With these substitutions, equation (2.26) becomes

$$R_{adj}^2 = 1 - \frac{df_n}{df_h}(1 - R^2). \tag{2.27}$$

In comparison, within an SEM context, let $df_n = (c - p) = (c - 0) = c$ in equation (2.24) denote the degrees of freedom associated with not having specified a CFA (or more general structural equation) model; that is, a situation in which no model parameters need to be estimated. The AGFI now can be expressed as

$$AGFI = 1 - \frac{df_n}{df_h}(1 - GFI). \tag{2.28}$$

Equations (2.27) and (2.28) should make the parallel between the AGFI and the R_{adj}^2 obvious: Both are shrunken versions of their biased, unadjusted counterparts, adjusted downward by a ratio comparing the degrees of freedom associated with not having specified a model versus those associated with the a priori hypothesized model of interest. Note, however, that, while the R_{adj}^2 can be used for inferential judgments regarding the proportion of explained variability in a regression model, the AGFI does not remove the statistical bias from the GFI completely; a component involving the sample size needs to be added to the shrinkage factor; see Maiti and Mukherjee (1990).

In Table 2.3, the GFI and AGFI for the SES example are reported as 0.999 and 0.994, respectively. Using either equation (2.24) or equation (2.28), the AGFI can be computed from the GFI by

$$AGFI = 1 - \frac{c}{df_h}(1 - GFI) = 1 - \frac{df_n}{df_h}(1 - GFI)$$

$$= 1 - \frac{10}{2}(1 - 0.999) = 0.995.$$

These values indicate that the observed data certainly fit the hypothesized model better than no model at all; both indices are very close to 1.00 and, thus, should indicate a good overall data-model fit.

The Normed and Nonnormed Fit Index and Their Extensions

Another class of fit measures are Bentler and Bonett's (1980) Normed and Nonnormed Fit Indices (NFI and NNFI, respectively) and their recently extended versions—the normed Comparative Fit Index (CFI) and the nonnormed Fit Index (FI) (Bentler, 1990). In a sense, these measures are related to the GFI and AGFI since they too approach the fit issue in a comparative way. However, rather than assessing the data-model fit for a hypothesized model by comparing it to not having a model at all—as is the case for the GFI and AGFI—Bentler and Bonett (1980) and Bentler (1990) took the approach of assessing fit by comparing a hypothesized model to some reasonable but more restrictive (fewer parameters to be estimated) *baseline model*.

Before explicitly defining the NFI and NNFI, it is worthwhile to briefly explain the general notion of *nested models* and a way of statistically comparing them. A specific model, M_j, is said to be nested within a less restricted model, say M_k, if it is a special case of the second model, that is, if M_j can be specified from M_k by fixing certain free parameters in M_k to 0. Model M_j is a more restricted version of M_k in the sense that M_j has the same structure as M_k, just with more parameters fixed to 0. On a continuum from most to least restricted, the *independence* and *saturated* models can be viewed as marking the two endpoints: The *independence model*, M_i, is a very restricted structure in which (a) no latent factors are hypothesized to underlie the observed variables (i.e., $\Lambda_X = I$ is specified to be the fixed identity matrix); (b) observed variables are measured without error (i.e., $\Theta_\delta = 0$ is set to the fixed 0 matrix); and (c) the observed variables are specified to be independent (i.e., Φ is specified to be a free but diagonal matrix). On the other end of the continuum lies the least restricted, just-identified, or *saturated model*, M_s, that includes as many free parameters as there are variances and covariances of observed variables, i.e., $df_s = (c - p) = 0$. The relationships between the models M_i, M_j, M_k, and M_s are shown schematically in Figure 2.4 (the role of another model in the figure, M_h, is clarified below).

Note from Figure 2.4 that more restricted models have higher overall χ^2-values than less restricted models. Since the former are nested within the latter, their goodness-of-fit can be compared statistically with differences in χ^2-tests: Suppose the sample size is large enough and models M_j and M_k in Figure 2.4 have asymptotic chi-square values of χ_j^2 and χ_k^2 with associated degrees of freedom df_j and df_k, respectively. Then $\chi^2 = (\chi_j^2 - \chi_k^2)$ also is distributed asymptotically as a chi-square with $df = (df_j - df_k)$ degrees of freedom. If this difference in chi-square value is statistically significant, then the data-model fit—as assessed by the overall chi-square statistic—for the less restricted model M_k is significantly better than for the more restricted model, M_j; otherwise, there is no statistically significant difference in the data-model fit for the two models and—all other things being equal—the more restricted model M_j should be chosen as a "better," more parsimonious representation of the structure underlying the observed variables. Although the above explanations seem to imply that two separate analyses are necessary to statistically compare the goodness-of-fit of two nested models, in practice, these comparisons can be conducted within a single analysis; see the discussion on the Lagrange Multiplier and Wald tests in the EQS example below.

Returning to the discussion of the fit indices NFI and NNFI, Bentler and Bonett (1980) conceptualized the fit issue in a model-comparative sense by assessing the data-model fit of a hypothesized structure M_h in comparison to

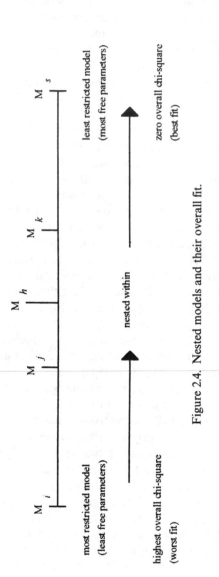

Figure 2.4. Nested models and their overall fit.

a more restricted baseline model, say M_j (see Figure 2.4). While any substantively reasonable model M_j could serve as a baseline (as long as it is nested within the hypothesized model M_h), it is usual practice to choose the independence model M_i as the standard for comparison (again, see Figure 2.4). Let F_h denote the minimum value of the fitting function F for the hypothesized model with associated degrees of freedom df_h. Similarly, let F_i be the minimum value of F for the independence baseline model with degrees of freedom df_i. Then the NFI can be defined as

$$\text{NFI} = \frac{\chi_i^2 - \chi_h^2}{\chi_i^2} = \frac{F_i - F_h}{F_i} = 1 - \frac{F_h}{F_i} \tag{2.29}$$

since $\chi^2 = (n - 1)F$ and the term $(n - 1)$ can be canceled from numerator and denominator. Furthermore, the NNFI is given by

$$\text{NNFI} = \frac{(\chi_i^2/df_i) - (\chi_h^2/df_h)}{\chi_i^2/df_i - 1} = \frac{(F_i/df_i) - (F_h/df_h)}{(F_i/df_i) - (1/(n - 1))}. \tag{2.30}$$

A comparison of the right-hand side expression of the NFI in equation (2.29) and the definition of the GFI in equation (2.23) makes the indices' similarities clear: Both are comparative in nature but deviate in the standard, to which a hypothesized model is compared (for the NFI this standard is a baseline model—usually the independence model—while for the GFI the standard is no model at all). The NFI is normed to a 0 to 1 range with larger values indicating a better data-model fit.

One disadvantage of the NFI is that it is affected by sample size and might not equal 1.00 even if the hypothesized model is correct (Bentler, 1990). To rectify this problem, Bentler and Bonett (1980) extended work by Tucker and Lewis (1973) and proposed their Nonnormed Fit Index (NNFI) given in equation (2.30). The NNFI, although not normed (i.e., values sometimes can be outside the 0 to 1 range), takes the degrees of freedom of both the baseline and the hypothesized models into account. As with the AGFI in equation (2.24), assuming a fixed baseline such as the independence model, more complex (less restrictive) hypothesized structures are penalized by a downward adjustment—while more parsimonious (more restrictive) models are rewarded by an increase—in the NNFI.

More recently, Bentler (1990) proposed two comparative fit indices that were designed to partially solve two major problems of the NFI and NNFI: (1) The normed Comparative Fit Index (CFI) was developed to address the above-mentioned fact that the NFI often underestimates data-model fit, especially for small samples; and (2) the nonnormed Fit Index (FI) was introduced to reduce the number of times NNFI values lie outside the 0 to 1 range. In addition, and contrary to other descriptive comparative fit indices, both the CFI and FI have a known associated population parameter, although their theoretical sampling distributions are, as of yet, unknown.

Let $l_h = [(n - 1)F_h - df_h]$ and $l_i = [(n - 1)F_i - df_i]$, where the terms on the right side of the equations are defined as above. Then the CFI and FI are given by

$$CFI = 1 - \frac{l_1}{l_2},$$ (2.31)

where $l_1 = \max(l_h, 0)$ and $l_2 = \max(l_h, l_i, 0)$; and

$$FI = 1 - \frac{l_h}{l_i}.$$ (2.32)

Clearly, the CFI is restricted to a 0 to 1 range while the FI is not [the numerator cannot be larger than the denominator in equation (2.31), but it could exceed the denominator in equation (2.32)]. Comparing the sets {CFI, FI} and {NFI, NNFI}, Bentler (1990) provided evidence that (a) the underestimation of fit happens less with the CFI than with the NFI; and (b) the FI behaves "better" than the NNFI at the endpoints of the [0, 1] interval; that is, "... FI will not be negative as frequently as NNFI.... [and] it will exceed 1.0 by a smaller amount [than the NNFI]" (p. 241).

For the SES example, the partial LISREL output in Table 2.3 lists the NFI, NNFI, and CFI—using the independence model as the baseline—as 0.997, 0.994, and 0.998, respectively [verification of these values using Table 2.3 and equations (2.29), (2.30), and (2.31), respectively, are left as Exercise 2.3]. Since in practice values above 0.90 for the indices often are considered to be indicative of good overall fit, one can conclude that the observed data indeed fit the model in Figure 2.1 better than the independence model.

The Parsimonious Fit Indices

Yet another class of fit indices is comprised of various "parsimony-based" measures that take the complexity of the hypothesized model into account when assessing overall data-model fit. Two early *parsimonious fit indices* are introduced below; consult Williams and Holahan (1994) for a complete review.

James *et al.* (1982) and Mulaik *et al.* (1989) criticized Jöreskog and Sörbom's GFI and AGFI and Bentler and Bonett's NFI by arguing that, even though researchers should abide by the principle of parsimony, they often opt for more complex (less restrictive) models since traditional measures of fit were designed to improve as the number of free parameters increased and the degrees of freedom decreased (see Figure 2.4). Each parameter freed removes a model-induced constraint on the unrestricted variance/covariance matrix of observed variables S and, thus, necessarily improves the fit between the data (represented by S) and the model [represented by the implied variance/covariance matrix, $\Sigma(\theta)$]. The authors proposed the Parsimony Goodness-of-Fit Index (PGFI) and the Parsimony Normed Fit Index (PNFI) "... to combine two logically interdependent pieces of information about a model, the goodness of fit of the model and the parsimony of the model, into a single index that gives a more realistic assessment of how well the model has been subjected to tests against available data and passed those tests" (Mulaik *et al.*, 1989, p. 439). The PGFI and PNFI are given by

$$\text{PGFI} = \frac{df_h}{df_n}\text{GFI} \qquad (2.33)$$

and

$$\text{PNFI} = \frac{df_h}{df_i}\text{NFI}, \qquad (2.34)$$

where df_h denotes the degrees of freedom associated with the hypothesized model, $df_n = c$ are the degrees of freedom when no model has been hypothesized (the number of nonredundant variances and covariances of the observed variables), and df_i are the degrees of freedom for the independence baseline model.

Comparing the AGFI in equation (2.28) with the PGFI in equation (2.33) reveals that both adjust the GFI downward for more complex, less parsimonious models, but they do so with different adjustment coefficients. Tanaka (1993, p. 21) showed that the PGFI provides a harsher penalty than the AGFI for hypothesizing a less restrictive model (with more parameters to estimate) rather than a more restrictive and parsimonious model. Similarly, the NNFI [equation (2.30)] and the PNFI [equation (2.34)] both adjust the NFI downward for more complex models, but in different ways.

For the confirmatory factor analysis of the model in Figure 2.1, the values of the PGFI and PNFI are given in Table 2.3 as 0.200 and 0.332, respectively. Of course, they can be computed directly by using equations (2.33) and (2.34): Since $df_h = (c - p) = (10 - 8) = 2$, $df_n = c = 10$, and $df_i = (10 - 4) = 6$, the Parsimonious Goodness-of-Fit Index is given by

$$\text{PGFI} = \frac{df_h}{df_n}\text{GFI} = \frac{2}{10}(0.999) = 0.200,$$

while the Parsimonious Normed Fit Index is

$$\text{PNFI} = \frac{df_h}{df_i}\text{NFI} = \frac{2}{6}(0.997) = 0.332.$$

With only $df_h = 2$ degrees of freedom, the model in Figure 2.1 is relatively "complex" in the sense that many parameters are being estimated ($p = 8$) relative to the number of available variances and covariances ($c = 10$). Thus, even though the GFI and NFI are very high (0.999 and 0.997, respectively; see Table 2.3), when penalizing for model complexity as suggested by James *et al.* (1982) and Mulaik *et al.* (1989), the indices of data-model fit are reduced greatly. From this perspective, the model of Figure 2.1 probably should be modified by imposing more restrictions on the observed variance/covariance matrix. Following brief concluding remarks regarding the data-model fit indices presented in the above sections, the questions of if and when model modifications are appropriate and justifiable are addressed.

Conclusion

The χ^2 and χ^2/df "badness" of fit measures, Jöreskog and Sörbom's goodness-of-fit indices, Bentler and Bonett's and Bentler's comparative fit indices, and James et al.'s and Mulaik et al.'s parsimonious fit indices are but a few examples of available measures of overall fit (for comprehensive reviews, see Marsh et al., 1988, and Tanaka, 1993). Due to a dependency on sample size, most—if not all—of these measures cannot have absolute cutoff points, so fit judgments must remain somewhat subjective (Marsh et al., 1988, p. 406). Also, since the above indices all depend on the multivariate normality assumption, results should be interpreted with caution if nonnormal data are analyzed. As should be evident from the above presentation, considerable controversies exist among experts on how best to define and measure data-model fit or misfit. Many, however, agree on a set of guiding principles as outlined by Bollen and Long (1993, pp. 6–8) and summarized as follows:

1. Use a strong underlying theory as the primary guide to assessing model adequacy since the objective of structural equation modeling in general (and confirmatory factor analysis in particular) is to understand a substantive area.
2. If appropriate and possible, formulate several alternative models prior to data analysis rather than considering a single model.
3. Whenever possible, compare fit results to ones from prior studies of the same or similar models.
4. In addition to measures assessing the fit for the hypothesized model as a whole, consider measures of fit of various components of a model.
5. Rely on multiple—rather than a single—measures of fit that represent different families of measures.
6. Use overall fit indices that take the degrees of freedom of a model into account and that depend as little as possible on sample size.

Using the latter two guidelines to summarize data-model fit results from the LISREL analysis of the CFA model in Figure 2.1 (see Table 2.3), one can make three overall observations:

1. Two classes of fit indices indicated good overall data-model fit (GFI = 0.999, AGFI = 0.994; and NFI = 0.997, NNFI = 0.994, CFI = 0.998); while
2. two classes of indices give some indications of possible data-model inconsistencies (χ^2 = 7.405 with p = 0.0247, χ^2/df = 3.703; and PGFI = 0.200, PNFI = 0.332); and

3. of the measures given in observations 1 and 2, less emphasis should be placed on the calculated values for the GFI, NFI, and the χ^2-statistic.

Recall that an inspection of individual parameter estimates in Table 2.2 revealed no obvious mis-specifications, with the exception, perhaps, that one of the indicator variables of parents' socioeconomic status had a somewhat low reliability estimate ($R^2 = 0.386$ for the observed variable *PaJntInc*). Given the above observations and the mainly illustrative purpose of the model in Figure 2.1 when viewed in isolation (the model is used as part of a larger, more general structural equation model in Chapter 3), we concluded that the data-model fit results from the LISREL analysis seem, overall, satisfactory.

Model Modification

After an initially hypothesized CFA or more general structural equation model has been specified and estimated, and fit results indicate no important data-model inconsistencies (as was the case for the model analyzed in the preceding LISREL example), the researcher probably has identified *a* possible structure that could have given rise to the data set under investigation. It is important to remember that this does *not* imply that the currently hypothesized model is *the* true underlying structure; *an analysis of the same data but with a different hypothesized model could have yielded the same fit results!*

If, however, any of the fit indices discussed in the previous sections indicate possible data-model inconsistencies (as in the EQS example discussed below), the researcher basically has three options on how to proceed:

1. In a strictly disconfirmatory sense, the presently specified model (the hypothesized theory) is rejected as being a possible representation of the structure underlying the data. Now a new model (new theory) must be conceptualized and developed, different data must be collected, and the new structure's viability must be evaluated via a second SEM analysis.
2. If additional, competing models were specified a priori, their fit results could be obtained from analyses of the same data set and then compared among themselves and to the ones from the initial model. If no important data-model inconsistencies are identified for an alternative model, that structure could be reported as *a* viable alternative representation of the underlying structure that generated the data. If more than one alternative model can be identified with acceptable fit results, the structure with the least severe data-model inconsistencies (and/or best cross-validation results; see below) is chosen as the most viable among competing models.
3. Information gained from the fit assessment and from other diagnostic statistics introduced below sometimes can be used to assist the researcher in modifying the initially hypothesized model, leading to a new proposed underlying structure with usually improved data-model fit.

Options 1 and 2 are scientifically sound and justifiable—albeit somewhat idealistic—actions that only few investigators opt to take. First, practical and logistic reasons usually prevent researchers from abandoning a carefully collected data set simply because it does not seem to fit an initially hypothesized structure. Second, with the availability of user-friendly and efficient SEM software packages, it sometimes becomes all too easy to succumb to impatience and analyze one favorite model before developing alternative structures. After all, why spend a lot of research time on concepualizing competing models when chances are good that the presently available data set fits the hypothesized model or a "slightly" modified version of the initial structure?

By far the most common approach to dealing with discovered data-model inconsistencies seems to be the one outlined in the third option: Fit results and other statistics are utilized in hopefully identifying and removing existing specification errors. Kaplan (1990) used the terms *external* and *internal specification error* to distinguish between errors committed by (a) having omitted important variables from inclusion in the model (external) and (b) having omitted important relations within the model (internal). Clearly, external specification errors can be identified only from a sound knowledge of one's substantive area. However, some empirical measures exist that can assist in identifying—and eliminating—internal specification errors. This is not to say that such measures by themselves provide enough justification to change an initially proposed model: On the contrary, *all* post-hoc modifications to a model *must* make substantive sense and be theoretically justifiable!

To locate possible internal specification errors, several researchers advocate the use of *modification indices* (MI) in conjunction with associated *expected parameter change statistics* (EPC). For each fixed parameter in a model, the MI is an estimate of the decrease in the χ^2 badness-of-fit measure if the parameter would be freed, and the EPC is an estimated value of that (now free) parameter. Saris, Satorra, and Sörbom (1987) and Kaplan (1990) suggested that (a) a fixed parameter with a large MI and EPC may be freed—provided, of course, this would make substantive sense; (b) a fixed parameter with a large MI but small EPC should remain fixed; and (c) a fixed parameter with a small MI and EPC should stay fixed. If a fixed parameter is associated with a small MI but large EPC, the situation is unclear and a more detailed examination of the source(s) of misspecification is recommended by the above authors.

Jöreskog and Sörbom (1993b) recommended a strategy for possible model modification that involves the observed χ^2-value and the modification indices: They argued that a large χ^2 (as compared to the degrees of freedom) should be taken as an indication that certain parameters should be freed to improve the data-model fit—and the modification indices provide an estimate of this improvement. On the other hand, a very small χ^2 might indicate that the model is *overfitted* in the sense that certain parameters could be fixed to 0 without significantly increasing the χ^2-value (within EQS, this can be tested via the Wald test; see the EQS example below).

Whatever modification strategy is used, two possible and important consequences need to be kept in mind. First, the two strategies above are not guaranteed to lead to the *one* model with the best data-model fit results. Several competing models might be generated that now must be compared on their substantive as well as statistical merits.

Second, realize that when the modified structure is reanalyzed and re-evaluated using the *same* data set that was utilized for the initial analysis, fit results usually *will* improve, not necessarily due to a truly "better" model (a structure that better reflects the "true" processes in the *population* that generated the data) but simply because a model has been fitted to a particular *sample* data set!

To clarify the latter point, recall that one of the main objectives of structural equation modeling is to assess the fit of the data—represented by the unrestricted variance/covariance matrix, Σ—to an a priori hypothesized model—represented by the model-implied variance/covariance matrix $\Sigma(\theta)$. Strictly speaking, the model is correct if and only if the hypothesis that $\Sigma = \Sigma(\theta)$ is true. Thus, a researcher who attempts to show that a sample data set fits a certain structure needs to verify that this hypothesis is a tenable one. However, if empirical fit measures indicate possible data-model inconsistencies, evidence exists that works against the researcher's goal. By modifying the hypothesized model (that is, changing the vector θ to, say, a vector θ^*) based on the sample-generated fit results, the investigator, in fact, modifies the hypothesis [from $\Sigma = \Sigma(\theta)$ to $\Sigma = \Sigma(\theta^*)$] after "peeking" at the data. Then, by reanalyzing the modified structure using the same data set, the researcher uses knowledge gained from the data and again tries to achieve the desired outcome, to retain the (now modified) hypothesis. Assuming that appropriate modifications were made, it should not be surprising at all that the sample data fit the new structure better than the initially hypothesized model; on the contrary, improved fit results are a necessary consequence of the modifications.

One way to address both of these possible consequences of post-hoc model modification (the generation of multiple competing models and the capitalization on chance by fitting a model to sample data) is to cross-validate the modified structure(s) with a new and independent sample. If such a sample is not available but the initial sample is large enough, Cudeck and Brown (1983) suggested mirroring cross-validation procedures available for linear regressions: Split the sample randomly into the calibration and validation subsamples and compute a *cross-validation index* (CVI) by measuring the distance between the unrestricted variance/covariance matrix obtained from the validation sample S_v and the model-implied variance/covariance matrix from the calibration sample $\Sigma(\hat{\theta}_c)$. That is, compute

$$CVI = F[S_v, \Sigma(\hat{\theta}_c)], \tag{2.35}$$

where F denotes some fitting function such as maximum likelihood or generalized least squares (see Chapter 3).

The smaller the value of the CVI, the better the estimated predictive validity of the model. However, what constitutes a "small" value for the CVI? If alternative models were hypothesized a priori or generated by following some model modification strategy, the one model with the smallest CVI could be chosen and reported as a viable structure representing the underlying process that generated the data. If no such competing models are available, one could compare the CVI value for the model under investigation to CVI results from (a) the saturated model (an unrestricted identified structure in which as many parameters are being estimated as there are variances and covariances of observed variables, i.e., $df = 0$), and/or (b) the independence model (the restricted model in which observed variables are hypothesized to be uncorrelated).

If for some reason it is not possible or reasonable to split the sample into calibration and validation subsamples (e.g., when the overall sample size is too small), Browne and Cudeck (1989) suggested using a single sample estimate of the *expected value of the cross-validation index* (ECVI). The ECVI is given by

$$\text{ECVI} = F[S, \Sigma(\hat{\theta})] + \frac{2p}{n-1}, \tag{2.36}$$

where $F[S, \Sigma(\hat{\theta})]$ is the minimum value of the fitting function for the hypothesized structure, p denotes the number of model-implied parameters to be estimated, and n is the sample size.

EQS EXAMPLE 2.1: A CFA OF ACADEMIC MOTIVATION, COLLEGE PRESTIGE, AND SES. For an illustration of confirmatory factor analysis with the EQS program—and of model modification—consider an analysis of the model shown in Figure 2.5, again based on data from the $n = 3094$ male participants in the ACE/UCLA study (see Appendix D for variable codes and summary statistics of all the observed variables in the model). Suppose the task is to estimate correlations among three constructs: respondent's academic motivation (*AcMotiv*), the prestige of the college the respondent attended (*ColgPres*), and respondent's socioeconomic status (*SES*). To this end, suppose further that (a) the observed indicator variables academic ability (*AcAbilty*), self-confidence (*SelfConf*), and degree aspirations (*DegreAsp*) are chosen to measure the latent factor *AcMotiv*; (b) the construct *ColgPres* is indicated by only one observed variable, namely college selectivity (*Selctvty*); and (c) the variables highest held academic degree (*Degree*), occupational prestige (*OcPrestg*), and current income (*Income*) are selected as indicators for *SES*. Note that, in Chapter 1, *DegreAsp*, *Selctvty*, and *Degree* were introduced as possible indicator variables of academic motivation, college prestige, and socioeconomic status and are the exogenous variables in the path analytical model of Figure 1.6. As with the previous LISREL example, the model in Figure 2.5 is not meant to represent the one true underlying structure of the observed variables; rather, it serves mainly illustrative purposes and is used in Chapter 3 to extend the path analytical model of Figure 1.6 to a full structural equation model with latent variables.

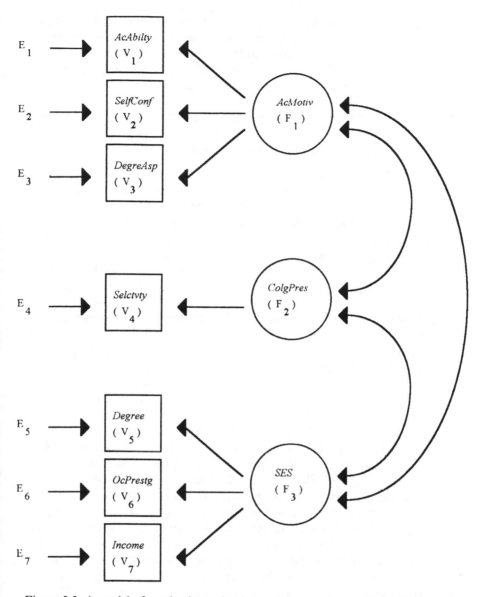

Figure 2.5. A model of academic motivation, college prestige, and SES in Bentler-Weeks notation.

Figure 2.5 also indicates the observed variables, associated measurement error terms, and the latent constructs in the Bentler-Weeks notation that is used in an EQS analysis of this model: As before, observed Variables are denoted by the letter V and Errors in observed variables are expressed by the letter E. Finally, the letter F designates the latent variables, or Factors.

EQS Input. The EQS input file is shown in Table 2.4. Much of the syntax was introduced in Chapter 1; in the following only new statements and other noteworthy aspects of the program are explained in detail.

Lines 1 through 4: The /TITLE and /SPECIFICATIONS sections should not require any explanation except to remind the reader that, since a correlation matrix is used as input (see lines 26 through 33), the statement "Matrix = Correlation;" must be included on line 4; otherwise, EQS would assume that the input matrix consists of variances and covariances. Also recall that, in general, all EQS statements end with a semicolon (;).

Lines 5 through 8: Note that the /LABELS section contains labels for the seven observed variables (lines 6 and 7) as well as for the three latent constructs (line 8).

Lines 9 through 16: Recall that an asterisk (*) indicates a free parameter to be estimated by the program; on the other hand, the absence of an asterisk indicates a fixed parameter (by default fixed to 1.00 unless another value is specified). Thus, the structural equations for the observed variables V1, V4, and V5 on lines 10, 13, and 14, respectively, imply that these variables serve as reference variables for the latent constructs. That is, by fixing the structural coefficient from the latent construct F1 to the observed variable V1— and from F2 to V4 and F3 to V5—to 1.00, the units of measurement for the latent variables *AcMotiv*, *ColgPres*, and *SES* are specified to equal those of the indicator variables *AcAbilty*, *Selctvty*, and *Degree*, respectively.

Lines 17 through 25: The /VARIANCES and /COVARIANCES sections must include the free variances and covariances of independent variables. Omitted variances or covariances among independent variables are fixed to 0 by default. For CFA models under the Bentler-Weeks model, the latent variables, F, and the measurement errors, E, are considered independent since they are specified to influence the dependent variables, V. Thus, on lines 18 through 21, all variances of the independent variables are specified to be free parameters except for the variance of the measurement error, E4, associated with the variable V4. Since the observed variable *Selctvty* (V4) is the only indicator of the latent construct *ColgPres*, it is assumed to be measured without error (see the section on identification above). Similarly, lines 23 through 25 specify covariances among the latent constructs F1, F2, and F3 as free parameters, but, by omission, covariances among measurement errors are fixed to 0.

Lines 26 through 35: As before, the /MATRIX and /STANDARD DEVIATIONS sections contain the input data in the form of a correlation matrix and standard deviations of observed variables.

Table 2.4. EQS Input File for the Model in Figure 2.5

1	/TITLE
2	Example 2.1. CFA Analysis of the Model in Figure 2.5
3	/SPECIFICATIONS
4	Variables = 7; Cases = 3094; Matrix = Correlation;
5	/LABELS
6	V1 = AcAbilty; V2 = SelfConf; V3 = DegreAsp; V4 = Selctvty;
7	V5 = Degree; V6 = OcPrestg; V7 = Income;
8	F1 = AcMotiv; F2 = ColgPres; F3 = SES;
9	/EQUATIONS
10	V1 = F1 + E1;
11	V2 = *F1 + E2;
12	V3 = *F1 + E3;
13	V4 = F2 + E4;
14	V5 = F3 + E5;
15	V6 = *F3 + E6;
16	V7 = *F3 + E7;
17	/VARIANCES
18	E1 to E3 = *;
19	E4 = 0;
20	E5 to E7 = *;
21	F1 to F3 = *;
22	/COVARIANCES
23	F3, F1 = *;
24	F3, F2 = *;
25	F2, F1 = *;
26	/MATRIX
27	1
28	.487 1
29	.236 .206 1
30	.382 .216 .214 1
31	.242 .179 .253 .254 1
32	.163 .090 .125 .155 .481 1
33	.064 .040 .025 .074 .106 .136 1
34	/STANDARD DEVIATIONS
35	.744 .782 1.014 1.990 .962 1.591 1.627
36	/LMTEST
37	Set = PEE;
38	/WTEST
39	/PRINT
40	Digit = 4;
41	/END

Lines 36 and 37: For the purpose of possible model modification, the /LMTEST section requests Lagrange Multiplier (LM) statistics—often called *modification indices;* see above—that are useful in judging the worth of including more free parameters in the model (making the structure more complex) in order to improve overall data-model fit. Asymptotically, LM statistics are equivalent to differences in χ^2-tests, comparing the nested hypothesized model M_h to one with more free parameters, say M_k (see Figure 2.4 and its explanation above). That is, they indicate and test the expected *decrease* in the overall χ^2-value when a presently fixed parameter (or set of parameters) is freed in a subsequent analysis. Associated with each LM value, EQS provides an expected parameter change statistic (EPC) that predicts the value of a fixed parameter if it is freed during a second analysis.

The "Set" subcommand on line 37 directs EQS to provide LM statistics for some specific fixed parameters. For the model in Figure 2.5, it could be of interest to assess whether freeing some of the fixed covariances among the measurement errors, E, would lead to a significant decrease in the overall χ^2-value. On the other hand, considering freeing the fixed structural coefficients linking the observed and latent variables would *not* make any substantive sense since the observed variables specifically were chosen as indicators of the particular latent variables as shown in Figure 2.5. (Note again that, within a traditional *exploratory* factor analysis framework, error terms are never allowed to covary and observed variables cannot be specified a priori to load on some underlying factors but not on others.) The statement "Set = PEE" requests LM tests for the fixed covariances among error terms— indicated by the two-letter combination EE—in the **PHI** matrix (*here, a variance/covariance matrix of the independent variables E and F in Bentler-Weeks notation*)—as indicated by the letter P preceding the EE combination.

Line 38: The /WTEST command requests Wald (W) statistics for free parameters in the model. In contrast to the Lagrange Multiplier values, these statistics test whether or not a presently *free* parameter (or set of parameters) could be fixed to 0 without significantly increasing the χ^2-value. Asymptotically, W statistics, too, are equivalent to differences in χ^2-tests, this time comparing the hypothesized model M_h to a nested one with a smaller number of free parameters, say, M_j (again, see Figure 2.4 and its explanation). They indicate and test the expected *increase* in the χ^2-value if a free parameter (or set of parameters) is fixed to 0; they can be used to assess whether or not the present model is overfitted and could be modified to a less complex, more parsimonious structure without eroding the overall data-model fit.

Lines 39 through 41: Finally, the /PRINT subcommand "Digit = 4;" on line 40 requests certain data-model fit indices—such as the GFI and AGFI, among others—in addition to the default measures NFI, NNFI, and CFI. The /END command on line 41 signals the end of the input file.

Selected EQS Output. Portions of the output from the confirmatory factor analysis of the model in Figure 2.5 that are pertinent to the data-model fit assessment and possible model modification are reproduced in Table 2.5. First, consider the set of overall fit indices: Whereas the χ^2-statistic for the hypothesized model of 155.501 ($df = 12, p < 0.001$) and the associated χ^2/df ratio of 12.958 indicate data-model inconsistencies (perhaps due to the large sample size), the various comparative fit indices (NFI = 0.946, NNFI = 0.913, CFI = 0.950) and the GFI and AGFI of 0.986 and 0.967, respectively, seem to indicate good overall data-model fit. Next, although not given in the EQS output, the parsimonious fit indices PGFI and PNFI can be computed easily from equations (2.33) and (2.34) as 0.423 and 0.541, respectively ($df_h = 12, df_n = 28$, and $df_i = 21$), giving a somewhat unclear message regarding the overall fit and parsimony of the model in Figure 2.5. What additional information is available that points to possible mis-specifications and that can assist in appropriately modifying the model, if necessary?

The χ^2-value for the independence model provided in the goodness-of-fit section of the output ($\chi_i^2 = 2905.519, df_i = 21$) and the minimum value of the fitting function F for the hypothesized model—obtained after five iterations (see Chapter 3)—given in the iterative summary section ($F[\mathbf{S}, \mathbf{\Sigma}(\hat{\boldsymbol{\theta}}_h)] = 0.05028$), can be used to compute the estimated expected value of the cross-validation index (ECVI) for the hypothesized, independence, and saturated models. First, using equation (2.36) yields an ECVI value for the hypothesized model of Figure 2.5 of

$$\text{ECVI}_h = F[\mathbf{S}, \mathbf{\Sigma}(\hat{\boldsymbol{\theta}}_h)] + \frac{2p_h}{n-1} = 0.05028 + \frac{2(16)}{3093} = 0.06063.$$

Second, since the reported chi-square for the independence model can be written as $\chi_i^2 = (n-1)F[\mathbf{S}, \mathbf{\Sigma}(\hat{\boldsymbol{\theta}}_i)]$ and the number of free parameters now is $p_i = 7$, the ECVI for the independence model is calculated as

$$\text{ECVI}_i = F[\mathbf{S}, \mathbf{\Sigma}(\hat{\boldsymbol{\theta}}_i)] + \frac{2p_i}{n-1} = \frac{\chi_i^2}{n-1} + \frac{2p_i}{n-1}$$

$$= \frac{2905.519}{3093} + \frac{2(7)}{3093} = 0.94391.$$

Finally, the ECVI for the saturated model is

$$\text{ECVI}_s = F[\mathbf{S}, \mathbf{\Sigma}(\hat{\boldsymbol{\theta}}_s)] + \frac{2p_s}{n-1} = 0 + \frac{2p_s}{n-1} = \frac{2(28)}{3093} = 0.01811$$

since $F[\mathbf{S}, \mathbf{\Sigma}(\hat{\boldsymbol{\theta}}_s)] = 0$ for the just-identified saturated model and the number of free parameters is $p_s = 28$.

Comparing the ECVI value for the hypothesized model to the values obtained for the independence and saturated models shows that ECVI$_s$ < ECVI$_h$ < ECVI$_i$, which leads to the conclusion that, technically, the saturated model has better predictive validity than the hypothesized model. However, Cudeck and Browne (1983) observed that this outcome is common in analyses based on large samples and that a "more parsimonious model

Table 2.5. Selected EQS Output from the CFA of the Model in Figure 2.5

GOODNESS OF FIT SUMMARY

INDEPENDENCE MODEL CHI-SQUARE = 2905.519 BASED ON 21 DEGREES OF FREEDOM

CHI-SQUARE = 155.501 BASED ON 12 DEGREES OF FREEDOM
PROBABILITY VALUE FOR THE CHI-SQUARE STATISTIC IS LESS THAN 0.001

BENTLER-BONETT NORMED	FIT INDEX = 0.946
BENTLER-BONETT NONNORMED	FIT INDEX = 0.913
COMPARATIVE	FIT INDEX = 0.950
LISREL GFI	FIT INDEX = 0.986
LISREL AGFI	FIT INDEX = 0.967

ITERATIVE SUMMARY

ITERATION	PARAMETER ABS CHANGE	ALPHA	FUNCTION
1	0.510612	1.00000	0.10090
2	0.102890	1.00000	0.05167
3	0.019630	1.00000	0.05031
4	0.003951	1.00000	0.05028
5	0.000988	1.00000	0.05028

STANDARDIZED SOLUTION:

$$ACABILTY = V1 = .8067 \ F1 + .5910 \ E1$$
$$SELFCONF = V2 = .5870*F1 + .8096 \ E2$$
$$DEGREASP = V3 = .3363*F1 + .9418 \ E3$$
$$SELCTVTY = V4 = 1.0000 \ F2 + .0000 \ E4$$
$$DEGREE = V5 = .8311 \ F3 + .5562 \ E5$$
$$OCPRESTG = V6 = .5779*F3 + .8161 \ E6$$
$$INCOME = V7 = .1517*F3 + .9884 \ E7$$

WALD TEST (FOR DROPPING PARAMETERS)
MULTIVARIATE WALD TEST BY SIMULTANEOUS PROCESS

CUMULATIVE MULTIVARIATE STATISTICS

STEP	PARAMETER	CHI-SQUARE	D.F.	PROBABILITY
************				

NONE OF THE FREE PARAMETERS IS DROPPED IN THIS PROCESS.

LAGRANGE MULTIPLIER TEST (FOR ADDING PARAMETERS)
ORDERED UNIVARIATE TEST STATISTICS:

					UNIVARIATE INCREMENT		
NO	CODE	PARAMETER	CHI-SQUARE	PROBABILITY	CHI-SQUARE	PROBABILITY	PARAMETER CHANGE
1	2 6	E2,E1	70.889	0.000			0.227
2	2 6	E5,E3	70.816	0.000			0.121
3	2 6	E3,E1	56.678	0.000			−0.122
4	2 6	E4,E2	35.679	0.000			−0.176
5	2 6	E7,E5	20.463	0.000			−0.158
6	2 6	E7,E6	15.158	0.000			0.171
7	2 6	E4,E1	14.497	0.000			0.145
8	2 6	E5,E1	13.610	0.000			−0.042
9	2 6	E4,E3	10.199	0.001			0.106
10	2 6	E6,E2	4.352	0.037			−0.036
11	2 6	E6,E5	2.695	0.101			0.365
12	2 6	E7,E4	2.179	0.140			0.078
13	2 6	E6,E4	1.536	0.215			−0.066
14	2 6	E7,E3	1.078	0.299			−0.029
15	2 6	E5,E4	0.688	0.407			0.038
16	2 6	E6,E1	0.685	0.408			0.014
17	2 6	E7,E1	0.685	0.408			0.015
18	2 6	E3,E2	0.628	0.428			0.011

Table 2.5 (cont.)

19	2 6	E6,E3	0.566	−0.018
20	2 6	E5,E2	0.519	0.008
21	2 6	E7,E2	0.005	−0.001
22	2 6	E4,E4	0.000	0.000
23	2 0	V1,F1	0.000	0.000
24	2 0	V4,F2	0.000	0.000
25	2 0	V5,F3	0.000	0.000

MULTIVARIATE LAGRANGE MULTIPLIER TEST BY SIMULTANEOUS
PROCESS IN STAGE 1
PARAMETER SETS (SUBMATRICES) ACTIVE AT THIS STAGE ARE:

PEE

CUMULATIVE MULTIVARIATE STATISTICS

					UNIVARIATE INCREMENT	
STEP	PARAMETER	CHI-SQUARE	D.F.	PROBABILITY	CHI-SQUARE	PROBABILITY
1	E2,E1	70.889	1	0.000	70.889	0.000
2	E5,E3	106.846	2	0.000	35.957	0.000
3	E7,E5	126.370	3	0.000	19.524	0.000
4	E4,E1	144.292	4	0.000	17.922	0.000
5	E6,E2	149.264	5	0.000	4.972	0.026

may ... still be judged as an adequate approximation to reality if the difference in cross-validation indices is small" (p. 153). In the example, $ECVI_h - ECVI_s = 0.04252$ seems small enough not to reject the model of Figure 2.5 based on its estimated cross-validation index.

Now consider the standardized solutions from which reliability estimates (the coefficients of determination for each of the structural equations) for the observed variables can be computed. As in Chapter 1, using the squared standardized coefficient estimates associated with the measurement errors E1 through E7 and subtracting these from 1.00, one obtains the reliability estimates shown in Table 2.6.

Note the low coefficients of determination for the variables *DegreAsp* ($R^2 = 0.113$) and *Income* ($R^2 = 0.023$). A possible explanation for the low reliability of *DegreAsp* could be the fact that all of the respondents were college freshmen and probably did not have a very clear idea regarding their future academic plans at the time of data collection. A reason for the low reliability estimate for the variable *Income* does not seem immediately apparent. Note, however, that while the two *SES* indicators *Degree* and *OcPrestg* are moderately correlated (0.481; see Appendix D), their correlations with *Income* are relatively low (0.106 and 0.136, respectively), possibly due to the ordinal measurement of the latter variable in the present data set. That is, self-reported highest held academic degree and occupational prestige here sccm to measure a different construct than reported annual income. Whatever the reasons for the low reliability estimate, it seems justified to modify the model in Figure 2.5 by removing the variable *Income* (V7) as an indicator of the latent construct *SES*.

Finally, review results from the Wald (W) and Lagrange Multiplier (LM) tests (consult the EQS manual, Chapter 6, for more in-depth explanations of the univariate and multivariate W and LM tests than can be given here). The part of the output relating to the requested Wald tests indicates that there are no free parameters in the model that could be fixed to 0 without significantly eroding the overall fit, that is, without significantly increasing the (already large and significant) overall χ^2-statistic. On the other hand, LM test results

Table 2.6. Estimated Reliability Coefficients
for Variables in the Model of Figure 2.5

Variable	Reliability Estimate (R^2)
AcAbilty (V1)	0.651
SelfConf (V2)	0.345
DegreAsp (V3)	0.113
Selctvty (V4)	1.000
Degree (V5)	0.691
OcPrestg (V6)	0.334
Income (V7)	0.023

Table 2.7. Modification Indices and Expected Parameter Change Statistics for Fixed Parameters with Significant Multivariate LM Test Results

Parameter	Label	Modification Index	EPC
E2, E1	*Self Conf, AcAbilty*	70.889	0.227
E5, E3	*Degree, DegreAsp*	70.816	0.121
E7, E5	*Income, Degree*	20.463	−0.158
E4, E1	*Selctvty, AcAbilty*	14.497	0.145
E6, E2	*OcPrestg, SelfConf*	4.352	−0.036

show that when certain measurement error terms are allowed to covary, the overall fit—as measured by the χ^2-statistic—improves significantly: Table 2.7 identifies those fixed measurement errors that the multivariate LM test suggests should be free to covary in order to improve overall data-model fit. Also, for each pair of parameters listed, the table identifies the associated univariate LM statistic (i.e., the modification index) and expected parameter change statistic (EPC) from the univariate LM test results in the EQS output.

Following Kaplan's (1990) model modification guidelines that were discussed previously, the first two parameters listed in Table 2.7 possibly could be freed during a second analysis: First, it seems theoretically justifiable to hypothesize a non-zero covariance between the measurement errors of (a) the indicators *Self Conf* and *AcAbilty* and (b) the variables *Degree* and *DegreAsp*, since, in both cases, unidentified common and/or correlated factors probably existed that contributed to the observed scores on the variables. Second, both parameters have relatively large modification indices—as compared to the overall chi-square value of 155.501—and associated sizable expected parameter change statistics.

The last parameter listed (the covariance between error terms of *OcPrestg* and *Self Conf*) probably should remain fixed to 0 due its relatively small MI and small EPC. Finally, there is no need to address the question of whether or not to free the remaining two parameters identified in Table 2.7 since (a) it was decided to remove the variable *Income* (V7) from the model based on its low reliability estimate and (b) the indicator variable *Selctvty* (V4) was assumed to be measured *without* error, and, thus, the covariance between E4 and E1 must remain fixed at 0.

In summary, inconclusive overall data-model fit results, Lagrange Multiplier tests, and reliability estimates seem to indicate that the model in Figure 2.5 possibly includes specification errors. Specifically, based on high modification indices and sizable expected parameter change statistics, possible internal specification errors are removed by modifying the initial structure so that the measurement error terms of (a) academic ability and self-confidence and (b) degree aspirations and highest held degree are free to covary. In addition, due to a very low reliability estimate, the indicator variable current income is removed from the model. Table 2.8 presents a partial output from an analysis of the so-modified structure (it is left to the reader to modify the EQS input file of Table 2.4 to reflect the appropriate changes; see Exercise 2.7).

First, a comparison of fit results in Table 2.8, and the ones shown in Table 2.5 show that, as expected, the data set fits the modified structure better than the initially hypothesized model in Figure 2.5. For example, the χ^2/df ratio of $27.464/5 = 5.493$ for the modified model is considerably smaller than the corresponding ratio from the previous analysis ($155.501/12 = 12.958$). Similarly, the values of the NFI, NNFI, CFI, GFI, and AGFI all are higher in Table 2.8 than in Table 2.5, indicating a better overall fit of the modified model than the initially proposed structure. Also, using equation (2.36) and the minimum value of the fitting function given in the ITERATIVE SUMMARY section of the output to calculate the ECVI for the modified model yields

$$\text{ECVI} = F[\mathbf{S}, \boldsymbol{\Sigma}(\hat{\boldsymbol{\theta}})] + \frac{2p}{n-1} = 0.00888 + \frac{2(16)}{3093} = 0.01923.$$

Since this value is lower than the one computed during the analysis of the original structure (there, $\text{ECVI} = 0.06063$), the present model has higher predictive validity than the model in Figure 2.5. On the other hand, note that the modifications made to the initial model resulted in a more "complex" structure in the sense that parsimonious fit indices are lower here than in the previous analysis: For the modified model, $\text{PGFI} = 0.237$ and $\text{PNFI} = 0.330$ [as can be verified by using equations (2.33) and (2.34) with $df_h = 5$, $df_n = 21$, and $df_i = 15$] whereas the corresponding values from the previous analysis were 0.541 and 0.423, respectively.

Second, the output in Table 2.8 shows that the three estimated structural parameters linking the observed variables *SelfConf*, *DegreAsp*, and *OcPrestg* to their corresponding latent constructs academic motivation (F1) and socio-economic status (F3)—as well as the estimated variances and covariances of the measurement error terms of (a) *SelfConf* and *AcAbility* and (b) *Degree* and *DegreAsp*—all are statistically significant. Further, the structural equations in standardized form indicate that, for the present sample of $n = 3094$ respondents, (a) the variable *AcAbilty* is a stronger indicator of the latent factor *academic motivation* (F1) than are *SelfConf* and *DegreAsp*, and (b) the observed variable *Degree* is a stronger indicator of socioeconomic status (F3) than is *OcPrestg*.

Third, and finally, the partial output shows the estimated correlation among the three latent variables ranging from 0.2938 for the constructs *SES* and *ColgPres* to 0.5773 for the factors *ColgPres* and *AcMotiv*. Again, note that these correlations are considerably higher than the corresponding (attenuated) correlations among observed variables. For example, while the correlation between the latent variables *AcMotiv* and *SES* is estimated at 0.4547, the correlations between the corresponding indicator variables only range between 0.090 and 0.242 (see the EQS input file in Table 2.4 or Appendix D).

Table 2.8. Partial EQS Output from an Analysis of the Modified Model of Figure 2.5

GOODNESS OF FIT SUMMARY

INDEPENDENCE MODEL
   CHI-SQUARE =   2832.727 BASED ON   15 DEGREES OF FREEDOM

CHI-SQUARE =   27.464 BASED ON   5 DEGREES OF FREEDOM
PROBABILITY VALUE FOR THE CHI-SQUARE STATISTIC IS LESS THAN 0.001

BENTLER-BONETT NORMED	FIT INDEX = 0.990
BENTLER-BONETT NONNORMED	FIT INDEX = 0.976
COMPARATIVE	FIT INDEX = 0.992
LISREL GFI	FIT INDEX = 0.997
LISREL AGFI	FIT INDEX = 0.988

ITERATIVE SUMMARY

ITERATION	PARAMETER ABS CHANGE	ALPHA	FUNCTION
1	0.484295	1.00000	0.37291
2	0.177397	1.00000	0.05030
3	0.037274	1.00000	0.00953
4	0.005306	1.00000	0.00890
5	0.001942	1.00000	0.00888
6	0.000334	1.00000	0.00888

MEASUREMENT EQUATIONS WITH STANDARD ERRORS AND TEST STATISTICS

ACABILTY = V1 =	1.0000 F1 +	1.0000 E1
SELFCONF = V2 =	.6830*F1 + .0389 17.5537	1.0000 E2
DEGREASP = V3 =	.8111*F1 + .0652 12.4402	1.0000 E3
SELCTVTY = V4 =	1.0000 F2 +	1.0000 E4
DEGREE = V5 =	1.0000 F3 +	1.0000 E5
OCPRESTG = V6 =	1.0661*F3 + .0856 12.4507	1.0000 E6

VARIANCES OF INDEPENDENT VARIABLES

$\underline{V}$	$\underline{F}$	
	I F1-ACMOTIV	.2301* I
	I	.0221 I
	I	10.4371 I
	I	I
	I F2-COLGPRES	3.9601* I
	I	.1007 I
	I	39.3256 I
	I	I
	I F3-SES	.6863* I
	I	.0569 I
	I	12.0619 I
	I	I

Table 2.8 (*cont.*)

VARIANCES OF INDEPENDENT VARIABLES

E			D
E1-ACABILTY	.3234*	I	I
	.0206	I	I
	15.7170	I	I
		I	I
E2-SELFCONF	.5042*	I	I
	.0174	I	I
	28.9984	I	I
		I	I
E3-DEGREASP	.8769*	I	I
	.0255	I	I
	34.4539	I	I
		I	I
E5-DEGREE	.2370*	I	I
	.0527	I	I
	4.4951	I	I
		I	I
E6-OCPRESTG	1.7513*	I	I
	.0748	I	I
	23.4249	I	I
		I	I

COVARIANCES AMONG INDEPENDENT VARIABLES

E			D
E2-SELFCONF	.1261*	I	I
E1-ACABILTY	.0156	I	I
	8.0696	I	I
		I	I
E5-DEGREE	.0896*	I	I
E3-DEGREASP	.0161	I	I
	5.5606	I	I
		I	I

STANDARDIZED SOLUTION:

ACABILTY = V1 =	.6448 F1 +	.7643 E1
SELFCONF = V2 =	.4190*F1 +	.9080 E2
DEGREASP = V3 =	.3837*F1 +	.9234 E3
SELCTVTY = V4 =	1.0000 F2 +	.0000 E4
DEGREE = V5 =	.8622 F3 +	.5067 E5
OCPRESTG = V6 =	.5551*F3 +	.8318 E6

Table 2.8 (*cont.*)

CORRELATIONS AMONG INDEPENDENT VARIABLES

V	F		
I	F2-COLGPRES	.5773*	I
I	F1-ACMOTIV		I
I			I
I	F3-SES	.4547*	I
I	F1-ACMOTIV		I
I			I
I	F3-SES	.2938*	I
I	F2-COLGPRES		I
I			I

Validity and Reliability from a CFA Perspective

Besides their function as a preparatory step in constructing the more general structural equation models discussed in Chapter 3, confirmatory factor analyses can be very useful in the evaluation of the validity and reliability of many types of instruments used in the social and behavioral sciences. In the previous LISREL and EQS examples, it already was shown how a coefficient of determination (R^2) obtained from a CFA can be used to estimate a variable's reliability. In many applications, confirmatory factor analysis is used for the sole purpose of assessing the psychometric properties of measures such as aptitude, achievement, or personality tests. Before illustrating this use of CFA by assessing the validity and reliability of a particular behavior assessment tool taken from the counseling psychology literature, the traditional definitions of validity and reliability are reviewed briefly and then refined from a CFA perspective (for thorough discussions of the reviewed concepts, consult the recommended references at the end of this chapter).

Validity

The *validity* of an instrument simply refers to the measure's overall property of indeed measuring what it was designed to measure. Traditionally, behavioral scientists have distinguished between content-related, criterion-related, and construct-related validity. An instrument's *content-related validity* refers to how well its items or scales cover the intended content (and associated cognitive processes the test takers are intended to use in responding to the items) and is assessed by a logical rather than statistical analysis. On the other hand, a measure's *criterion-related validity* is determined by computing a correlation coefficient, usually called the *validity coefficient*, between scores on the instrument to be validated and scores on some criterion measure that is used as a standard of comparison and is assumed to measure the same construct as the instrument in question. As Bollen (1989, p. 187) noted, however, this correlation does not only depend on how closely the instrument measures the construct it was designed to measure, but it also depends on the association between the criterion variable and the construct. Thus, an important shortcoming of the validity coefficient is that its value not only changes as a function of the error in the measure of interest, but also as a function of the error in the criterion measure.

Finally, to gather evidence of the *construct-related validity* of an instrument, a researcher determines the extent of agreement between observed and hypothesized correlations of the measure of interest and other measures of either the same or other constructs: While the instrument to be validated should correlate *highly* with measures of the same or strongly associated constructs, it should produce a *low* correlation with measures of different constructs that have very little association with the construct to be measured. Here, the assumption is made that correlations between observed measures accurately reflect relationships between underlying constructs. However, the former correlations do not only depend on the relationships between the constructs, but they also depend on the reliabilities of the observed measures (Bollen, 1989, p. 190). This can lead easily to false conclusions regarding a measure's construct validity: For example, the validity of an instrument could be seriously underestimated if the comparison instrument were very unreliable.

The validity definitions just reviewed have two major disadvantages in common. First, a given instrument, scale, or item is assumed to measure one underlying construct only; second, involved latent constructs are not incorporated explicitly into the operational validity definitions (i.e., only correlations between *observed* variables are used as evidence of validity). To address these shortcomings, exploratory factor analysis (EFA) sometimes is used to provide evidence of an instrument's construct-related validity. As was mentioned before, this techniques generates the constructs that might underly the observed variables by statistical—rather than logical—means. "Discovered" factors often are difficult to name and interpret since the observed variables (e.g., instrument scales or items) cannot be grouped a priori on substantive grounds but, instead, are related to *all* of the identified factors to varying degrees. Another problem with this approach to validity assessment is that often it is not clear from EFA results how many factors need to be retained for final interpretation, i.e., of the factors generated by the analysis, how many are meaningful and interpretable?

Approaching an instrument's validity from a confirmatory factor analysis perspective addresses all of the above-mentioned concerns. Based on substantive knowledge, CFA allows a researcher to a priori specify how a set of scales or items should be related theoretically to a set of underlying constructs. If CFA results indicate acceptable data-model fit (i.e., overall, the data failed to disconfirm the a priori hypothesized measurement structure of the instrument), the investigator can proceed with a more in-depth assessment of validity. For example, Bollen (1989) and others suggested estimating the validity of a latent construct's indicator by the magnitude of the factor loading linking the observed and latent variable [i.e., the appropriate structural coefficient in the Λ_x matrix of equation (2.5)]. Similar to the use of metric and standardized coefficients in a regression (see Chapter 1), unstandardized factor loadings can be used to compare validity results across different samples, while standardized structural coefficients should be utilized when comparing the validity of different observed variables based on data from the same sample (Bollen, 1989, pp. 198–200).

Other indications of an instrument's validity can be inferred from a CFA analysis. For example, Bollen (1989) provided an algorithm to calculate that part of the variability in an indicator variable which is uniquely attributable to an underlying construct that the variable was designed to measure: The more variance the latent construct explains, the higher the validity of the observed variable. Also, one could examine the extent of agreement between hypothesized and estimated relationships among *latent* constructs. If, for example, a test was designed to measure two or more different but related facets of personality (e.g., the thinking, feeling, and acting dimensions of behavior; see the LISREL and EQS examples below), estimated covariances between the corresponding latent variables in the Φ matrix should be sizable. On the other hand, a CFA analysis of a general academic achievement battery assessing a variety of relatively independent constructs (e.g., mathematics and foreign language skills) should result in fairly low correlations among the latent variables. Such correspondence between theoretically derived hypotheses and empirical results could be used as partial evidence of an instrument's validity.

Reliability

An instrument's *reliability* refers to how consistently the instrument measures whatever it was designed to measure. Usually, researchers distinguish between test–retest, parallel forms, and internal consistency reliability. *Test–retest reliability* assesses how stable measurements are over time and is estimated by the correlation (the *coefficient of stability*) between scores obtained from two administrations of the same instrument to the same respondents over some period of time. On the other hand, *parallel forms reliability* can be defined as a measure of the equivalence of two forms of the same test. If the correlation (the *coefficient of equivalence*) between two versions of the same instrument is high, evidence exists that they indeed are parallel. Due to the time lag between measurements, a disadvantage of both approaches (test–retest and parallel forms) is the potential presence of carry-over effects (practice, memory, attitude) that might lead to either an under- or overestimation of the reliability of a measure.

To overcome this difficulty, an instrument's *internal consistency* can be assessed that requires only a single administration for the computation of a reliability coefficient. There are various ways of assessing the internal consistency of an instrument. Probably the most straightforward way is to calculate the correlation between two parts of a test by splitting the total number of items into two halves of equal length. The problem here lies in the decision of how to split the test: Should one select half of the items at random, or perhaps chose the odd-numbered items for one-half of test and the even-numbered items for the other? Alternative approaches to assessing internal consistency that do not depend on the particular way the test is split into two halves include the computation of one of the Kuder-Richardson formulas (KR-20 or KR-21) or the Cronbach alpha coefficient.

Estimating an instrument's reliability from an internal consistency perspective largely illiminates the chances of carry-over effects. Note, however, that the reliability of single-item measures cannot be assessed. In addition, all traditional definitions of reliability do not allow for correlated measurement errors of items or scales (i.e., the off-diagonal elements in the Θ_δ matrix are assumed to be 0), and observed variables cannot be indicators for more than one underlying construct (for an example of when such a situation could occur, see the LISREL and EQS analyses below). Within a confirmatory factor analysis framework, reliability can be redefined to address the shortcomings just mentioned. The proportion of variance, R^2, in an observed variable that is accounted for by all latent constructs that are hypothesized to affect it has been proposed by Bollen (1989) and others as a way to estimate a measure's reliability. As was already illustrated in previous examples, this coefficient of determination is readily available from a LISREL analysis and easily computed from an EQS output.

The *Hutchins Behavior Inventory:* An Example

Before illustrating an actual confirmatory factor analysis for the purpose of validity and reliability assessment with LISREL and EQS, some background information should be given on the particular instrument used, which is taken from the field of counseling psychology. Of course, the test's manual (Hutchins and Mueller, 1992) gives more detailed information regarding the instrument's theoretical background, scale development, and scoring and should be consulted by interested readers.

Over the last few decades, several counseling professionals have moved toward incorporating cognitive, emotional, and psychomotor components into their views of human behavior and therapy approaches. Measurement tools have emerged that attempt to assess a person's degree of integration of the thinking (cognitive), feeling (emotional), and acting (psychomotor) components of behavior. That is, does the client approach a particular life situation with a primarily thinking, feeling, or acting orientation or, depending on the context, does the individual balance these three behavior components? One such measure is the *Hutchins Behavior Inventory* (*HBI*; Hutchins, 1992).

Briefly, the *HBI* provides two type of scales (comparison and characteristic) that can assist counselors in assessing (1) a person's integration of the thinking, feeling, and acting dimensions and (2) how characteristic each of the three dimensions is of a person's behavior. The *HBI*'s characteristic scales, T_c, F_c, and A_c, are Likert-type indicators of the intensity of the thinking, feeling, and acting dimensions. The comparison scales, T_f, F_a, and A_t, on the other hand, consider the constructs of thinking, feeling, and acting in a pairwise fashion and were designed to assess the dominance of one construct over another. For example, the T_f scale was developed to assess whether an individual in a particular situation behaves with a primarily thinking (cognitive)—as opposed to feeling (emotional)—orientation or if the person's approach is of a more balanced nature, expressing characteristics of both a cognitive and emotional response to the given situation. Similarly, the F_a and A_t scales were designed to assess the dominance of one construct over the other on a feeling–acting and acting–thinking continuum, respectively.

The hypothesized structure underlying the instrument's six scales is shown in Figure 2.6 (see the test's manual; Hutchins and Mueller, 1992, p. 31). The same model in general CFA notation (including that used in the Benter-Weeks model for later reference) is given in Figure 2.7.

Note that the model in Figure 2.6 contains six observed variables (the *HBI* scales T_f, T_c, F_a, F_c, A_t, and A_c), which are hypothesized to load in a particular way on three underlying constructs (the behavior components: thinking,

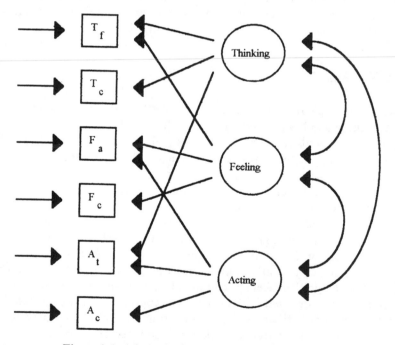

Figure 2.6. A hypothesized CFA model of the *HBI*.

feeling, and acting). Furthermore, the latent variables are hypothesized to covary, and each of the three comparison scales (T_f, F_a, and A_t) is used as an indicator of *two* underlying constructs. It is important to realize that indeed this CFA model is based on several strong theoretical rationales:

1. A rich body of literature, suggesting crucial links between human cognition, emotion, and overt behavior, led the *HBI*'s developer more than a decade prior to the instrument's release to operationally define human behavior to include "how a person thinks, feels, and acts" (Hutchins, 1979, p. 529). Subsequent theoretical writings (e.g., Hutchins, 1984; Hutchins and Cole, 1992) on the interplay between the cognitive, emotional, and psychomotor components provide a clear rationale for hypothesizing non-zero covariances between the latent constructs thinking, feeling, and acting in the model of Figure 2.6.
2. By design, the three comparison scales T_f, F_a, and A_t each should serve as indicators of two underlying constructs since the scales assess an individual's degree of integration of two behavior dimensions, thinking and feeling, feeling and acting, and acting and thinking, respectively.
3. The Likert-type characteristic scales T_c, F_c, and A_c were specifically developed to independently measure how characteristic each of the three

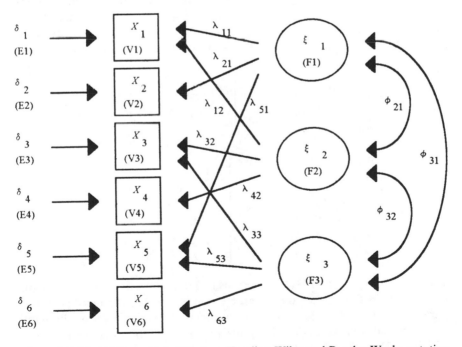

Figure 2.7. The *HBI* model in Jöreskog-Keesling-Wiley and Bentler-Weeks notation in parentheses.

behavior components is of a person's behavior in a particular situation. Thus, they should be understood as independent indicators of the individual latent variables of thinking, feeling, and acting.

4. Clearly, none of the *HBI*'s scales can be viewed as a perfect measure of any of the underlying constructs but, instead, must include measurement error. However, given the design of the instrument, the errors should *not* covary since (a) characteristic scales are independent of each other and of the comparisons scales, and (b) possible dependencies among comparison scales are incorporated in the model by specifying each such scale to be an indicator of two latent constructs.

As with any other CFA structure, the *HBI* model depicted in Figure 2.6 can be represented by the matrix equation $\mathbf{X} = \mathbf{\Lambda}_X \boldsymbol{\xi} + \boldsymbol{\delta}$. Specifically,

$$\begin{bmatrix} X_1 \\ X_2 \\ X_3 \\ X_4 \\ X_5 \\ X_6 \end{bmatrix} = \begin{bmatrix} \lambda_{11} & \lambda_{12} & 0 \\ \lambda_{21} & 0 & 0 \\ 0 & \lambda_{32} & \lambda_{33} \\ 0 & \lambda_{42} & 0 \\ \lambda_{51} & 0 & \lambda_{53} \\ 0 & 0 & \lambda_{63} \end{bmatrix} \begin{bmatrix} \xi_1 \\ \xi_2 \\ \xi_3 \end{bmatrix} + \begin{bmatrix} \delta_1 \\ \delta_2 \\ \delta_3 \\ \delta_4 \\ \delta_5 \\ \delta_6 \end{bmatrix}. \tag{2.37}$$

In addition, after assigning units of measurement to the three latent constructs of thinking (ξ_1), feeling (ξ_2), and acting (ξ_3), this time by standardizing them to unit variance, the variance/covariance matrix of the latent variables, $\mathbf{\Phi}$, and the variance/covariance matrix of error terms, $\mathbf{\Theta}_\delta$, can be written as

$$\mathbf{\Phi} = \begin{bmatrix} 1 & & \\ \phi_{21} & 1 & \\ \phi_{31} & \phi_{32} & 1 \end{bmatrix} = \begin{bmatrix} 1 & & \\ \sigma_{\xi_2\xi_1} & 1 & \\ \sigma_{\xi_3\xi_1} & \sigma_{\xi_3\xi_2} & 1 \end{bmatrix} \tag{2.38}$$

and

$$\mathbf{\Theta}_\delta = \begin{bmatrix} \theta_{11} & & & & & \\ 0 & \theta_{22} & & & & \\ 0 & 0 & \theta_{33} & & & \\ 0 & 0 & 0 & \theta_{44} & & \\ 0 & 0 & 0 & 0 & \theta_{55} & \\ 0 & 0 & 0 & 0 & 0 & \theta_{66} \end{bmatrix} = \begin{bmatrix} \sigma_{\delta_1}^2 & & & & & \\ 0 & \sigma_{\delta_2}^2 & & & & \\ 0 & 0 & \sigma_{\delta_3}^2 & & & \\ 0 & 0 & 0 & \sigma_{\delta_4}^2 & & \\ 0 & 0 & 0 & 0 & \sigma_{\delta_5}^2 & \\ 0 & 0 & 0 & 0 & 0 & \sigma_{\delta_6}^2 \end{bmatrix}. \tag{2.39}$$

The matrices $\mathbf{\Lambda}_X$, $\mathbf{\Phi}$, and $\mathbf{\Theta}_\delta$ in equations (2.37)–(2.39) completely specify the *HBI* model in Figure 2.6, and it can be seen easily that the number of free parameters, $p = 18$, is less than the number of available (nonredundant) variances and covariances, $c = 21$. Thus, the model probably is overidentified with $df = 21 - 18 = 3$.

Table 2.9. LISREL Input for an Analysis of the *HBI* Model
in Figure 2.6

1	Example 2.2. Validity and Reliability of the HBI
2	DA NI = 6 NO = 167
3	LA
4	Tf Tc Fa Fc At Ac
5	CM SY
6	.436
7	.045 .196
8	−.349 −.048 .468
9	−.145 .126 .112 .243
10	−.037 .013 −.117 .037 .284
11	.029 .165 −.112 .127 .100 .280
12	MO NX = 6 NK = 3 LX = FU,FI PH = SY,FR TD = DI
13	LK
14	Thinking Feeling Acting
15	FR LX(1,1) LX(2,1) LX(5,1)
16	FR LX(1,2) LX(3,2) LX(4,2)
17	FR LX(3,3) LX(5,3) LX(6,3)
18	OU SC ND = 3

LISREL EXAMPLE 2.2: THE VALIDITY AND RELIABILITY OF THE *HBI*

LISREL Input. The following analyis of the *HBI* model in Figure 2.6 is based on a data set containing *HBI* results from $n = 167$ residence-hall counselors at a large southeastern land-grant university (see Mueller *et al.*, 1990). An appropriate LISREL input file is provided in Table 2.9. Since no new features are introduced, the input should, for the most part, be self-explanatory. Note that, in this example, a symmetric variance/covariance matrix of the observed variables is used as input data, as is indicated by CM SY on line 5. The information on the MOdel line (line 12) and lines 15 through 17 specify the *HBI* model of Figure 2.6. Initially, the structural coefficient matrix Λ_x is specified to be a FUll and FIxed matrix, but subsequently appropriate elements are set free (see program lines 15 through 17 and Figure 2.7). Further, no reference variables are used to assign units of measurement to the latent variables thinking, feeling, and acting; thus, by default, LISREL assumes the diagonal elements of the Φ matrix (containing the variances and covariances of the latent variables) to be fixed to 1.0 even though Φ is specified to be a free (and symmetric) matrix on line 12. Finally, the statement TD = DI on the same program line indicates to LISREL that all diagonal elements in the Θ_δ matrix (the variance/covariance matrix of measurement errors) are free parameters with all off-diagonal elements fixed to 0.

Table 2.10. Partial LISREL Output from an Analysis of the *HBI* Model in Figure 2.6

LISREL ESTIMATES (MAXIMUM LIKELIHOOD)

LAMBDA-X

	Thinking	Feeling	Acting
Tf	0.762	−0.904	—
	(0.085)	(0.081)	
	8.955	−11.220	
Tc	0.418	—	—
	(0.028)		
	14.741		
Fa	—	0.852	−0.826
		(0.077)	(0.081)
		11.077	−10.201
Fc	—	0.413	—
		(0.032)	
		12.922	
At	−0.751	—	0.885
	(0.124)		(0.125)
	−6.059		7.089
Ac	—	—	0.448
			(0.035)
			12.953

PHI

	Thinking	Feeling	Acting
Thinking	1.000		
Feeling	0.721	1.000	
	(0.053)		
	13.636		
Acting	0.889	0.689	1.000
	(0.025)	(0.058)	
	36.221	11.882	

THETA-DELTA

Tf	Tc	Fa	Fc	At	Ac
0.031	0.021	0.029	0.072	0.118	0.080
(0.020)	(0.011)	(0.022)	(0.011)	(0.024)	(0.013)
1.577	2.006	1.333	6.851	4.875	6.172

SQUARED MULTIPLE CORRELATIONS FOR X-VARIABLES

Tf	Tc	Fa	Fc	At	Ac
0.929	0.892	0.938	0.702	0.583	0.716

Table 2.10 (*cont.*)

GOODNESS OF FIT STATISTICS

CHI-SQUARE WITH 3 DEGREES OF FREEDOM = 0.882 (P = 0.830)

MINIMUM FIT FUNCTION VALUE = 0.00531

EXPECTED CROSS-VALIDATION INDEX (ECVI) = 0.222
ECVI FOR SATURATED MODEL = 0.253
ECVI FOR INDEPENDENCE MODEL = 3.569

CHI-SQUARE FOR INDEPENDENCE MODEL WITH 15 DEGREES OF
FREEDOM = 580.464

GOODNESS OF FIT INDEX (GFI) = 0.998
ADJUSTED GOODNESS OF FIT INDEX (AGFI) = 0.988
PARSIMONY GOODNESS OF FIT INDEX (PGFI) = 0.143

NORMED FIT INDEX (NFI) = 0.998
NON-NORMED FIT INDEX (NNFI) = 1.019
PARSIMONY NORMED FIT INDEX (PNFI) = 0.200
COMPARATIVE FIT INDEX (CFI) = 1.000

COMPLETELY STANDARDIZED SOLUTION

LAMBDA-X

	Thinking	Feeling	Acting
Tf	1.154	− 1.369	—
Tc	0.945	—	—
Fa	—	1.246	−1.207
Fc	—	0.838	—
At	−1.409	—	1.661
Ac	—	—	0.846

PHI

	Thinking	Feeling	Acting
Thinking	1.000		
Feeling	0.721	1.000	
Acting	0.889	0.689	1.000

THETA-DELTA

Tf	Tc	Fa	Fc	At	Ac
0.071	0.108	0.062	0.298	0.417	0284

Selected LISREL Output. Partial results from the analysis of the *HBI* model is shown in Table 2.10. Perusing the output indicates that there are no unreasonable parameter estimates. The data-model fit indices give some initial overall evidence of the validity of the *HBI*: They suggest that the measurement structure proposed in Figure 2.6 is indeed a possible and plausible way of relating the six *HBI* scales to the thinking, feeling, and acting dimensions of behavior. For example, the low χ^2-value of 0.882 ($df = 3$, $p = 0.830$) —its nonsignificance due either to a "perfect" data-model fit or to a lack of statistical power (recall the small sample size of $n = 167$)—indicates no overall data-model inconsistencies. In fact, the model could be overfitted; Wald tests can be used to gauge whether some parameters could be fixed to 0 without significantly erroding the overall fit (see the EQS analysis below and, also, consider Exercise 2.2 at the end of the chapter).

All free structural parameters (the unstandardized validity coefficients) in the Λ_x matrix and the three covariances among the latent constructs are significant while the two variances of measurement errors associated with the comparison scales T_f and F_a are nonsignificant. In addition, comparing magnitudes of the standardized structural coefficients, it can be seen that the A_t

Table 2.11. EQS Input for an Analysis of the *HBI* Model in Figure 2.6

1	/TITLE
2	Example 2.2. Validity and Reliability of the HBI
3	/SPECIFICATIONS
4	Variables = 6; Cases = 167;
5	/LABELS
6	V1 = Tf; V2 = Tc; V3 = Fa; V4 = Fc; V5 = At; V6 = Ac;
7	F1 = Thinking; F2 = Feeling; F3 = Acting;
8	/EQUATIONS
9	V1 = *F1 + *F2 + E1;
10	V2 = *F1 + E2;
11	V3 = *F2 + *F3 + E3;
12	V4 = *F2 + E4;
13	V5 = *F1 + *F3 + E5;
14	V6 = *F3 + E6;
15	/VARIANCES
16	F1 to F3 = 1;
17	E1 to E6 = *;
18	/COVARIANCES
19	F3, F1 = *;
20	F3, F2 = *;
21	F2, F1 = *;
22	/MATRIX
23	.436
24	.045 .196
25	−.349 −.048 .468
26	−.145 .126 .112 .243
27	−.037 .013 −.117 .037 .284
28	.029 .165 −.112 .127 .100 .280
29	/WTEST
30	Apriori = (F3,F1), (F3,F2), (F2,F1);
31	/END

comparison scale seems to be the most valid indicator of both the thinking and acting constructs while for the feeling domain, the T_f scale might be the most valid indicator. However, the somewhat unusual situation that two of the three standardized structural coefficients in each of the three equations are larger than 1.00 indicates that conclusions should be drawn only with caution.

The estimated correlations between the latent constructs thinking, feeling, and acting (ranging from 0.721 to 0.889 in the standardized Φ matrix) mirror the theoretical hypothesis that cognitive, emotional, and action-oriented behavioral components are closely related and, thus, give additional initial evidence of the validity of the HBI for the group of $n = 167$ residence-hall counselors. Finally, the squared multiple correlations (R^2) for the six scales (ranging from a low of 0.583 for the A_t scale to a high of 0.938 for the F_a scale) indicate acceptable reliability of the HBI in the present sample. Note that the scales with the highest reliabilities (A_t and F_a) are, of course, the ones with the lowest measurement error variances (here, these variances are nonsignificant, as was mentioned earlier).

EQS EXAMPLE 2.2: THE VALIDITY AND RELIABILITY OF THE HBI

EQS Input. The program for an EQS analysis of the HBI data is presented in Table 2.11. It follows the same basic structure as the previous EQS example of a CFA (see Table 2.4) and thus should not require any explanation except to note that (a) since no reference variables are provided for the constructs thinking (F1), feeling (F2), and acting (F3) in the /EQUATIONS section (lines 8 through 14), the variances of the latent factors must be fixed to 1.0 in the /VARIANCES section (lines 15 through 17) to provide the latent variables with a unit of measurement; and (b) a Wald test (/WTEST) is requested on line 29 for the hypothesized non-zero associations between the latent variables thinking (F1), feeling (F2), and acting (F3). Again, EQS's Wald test option can be used to assess whether a model is overfitted, i.e., whether there are free parameters that could be fixed to 0 without significantly increasing the overall χ^2-statistic. Due to the particular design of the HBI, it only makes substantive sense to subject the free covariances between the latent variables (F1, F2, and F3) to the Wald test. Thus, the statement "Apriori = (F3,F1), (F3,F2), (F2,F1);" on line 30 requests the test first (a priori) for those associations; EQS tests other free parameters only if fixing the three specified parameters to 0 does not significantly increase the χ^2-statistic.

Selected EQS Output. A partial EQS output file for the HBI analysis is shown in Table 2.12; all results equal those from the LISREL analysis in Table 2.10 and were discussed previously. In addition, note that the re-

Table 2.12. Partial EQS Output from an Analysis of the *HBI* Model in Figure 2.6

GOODNESS OF FIT SUMMARY

CHI-SQUARE =   0.882 BASED ON   3 DEGREES OF FREEDOM
PROBABILITY VALUE FOR THE CHI-SQUARE STATISTIC IS 0.82981

BENTLER-BONETT NORMED   FIT INDEX = 0.998
BENTLER-BONETT NONNORMED   FIT INDEX = 1.019
COMPARATIVE   FIT INDEX = 1.000

MEASUREMENT EQUATIONS WITH STANDARD ERRORS AND TEST
STATISTICS

TF	= V1	=	.763*F1	−.904*F2	+1.000 E1
			.085	.081	
			8.954	−11.218	
TC	= V2	=	.418*F1	+1.000 E2	
			.028		
			14.738		
FA	= V3	=	.852*F2	−.826*F3	+1.000 E3
			.077	.081	
			11.076	−10.201	
FC	= V4	=	.413*F2	+1.000 E4	
			.032		
			12.922		
AT	= V5	=	−.751*F1	+.885*F3	+1.000 E5
			.124	.125	
			−6.060	7.090	
AC	= V6	=	.448*F3	+1.000 E6	
			.035		
			12.954		

VARIANCES OF INDEPENDENT VARIABLES

	E
E1-TF	.031*
	.020
	1.570
E2-TC	.021*
	.011
	2.010
E3-FA	.029*
	.022
	1.336
E4-FC	.072*
	.011
	6.851

Table 2.12 (*cont.*)

E5-AT	.118*
	.024
	4.877
E6-AC	.079*
	.013
	6.170

COVARIANCES AMONG INDEPENDENT VARIABLES

F

F2-FEELING	.721*
F1-THINKING	.053
	13.640
F3-ACTING	.889*
F1-THINKING	.025
	36.220
F3-ACTING	.689*
F2-FEELING	.058
	11.884

STANDARDIZED SOLUTION:

TF	=V1	=	1.155*F1	−1.370*F2	+.265 E1
TC	=V2	=	.944*F1	+.329 E2	
FA	=V3	=	1.246*F2	−1.207*F3	+.250 E3
FC	=V4	=	.838*F2	+.546 E4	
AT	=V5	=	−1.409*F1	+1.661*F3	+.646 E5
AC	=V6	=	.846*F3	+.533 E6	

WALD TEST (FOR DROPPING PARAMETERS)
MULTIVARIATE WALD TEST BY   APRIORI   PROCESS
CUMULATIVE MULTIVARIATE STATISTICS

STEP	PARAMETER	CHI-SQUARE	D.F.	PROBABILITY
1	F3,F2	141.225	1	0.000
2	F2,F1	187.494	2	0.000
3	F3,F1	1456.903	3	0.000

UNIVARIATE INCREMENT

CHI-SQUARE	PROBABILITY
141.225	0.000
46.268	0.000
1269.409	0.000

quested Wald tests indicate that fixing any of the covariances between the latent constructs thinking, feeling, and acting would significantly increase the χ^2-statistic. Thus, no modifications to the hypothesized model in Figure 2.6 should be considered.

Summary

Confirmatory factor analysis (CFA) is a method for evaluating a priori hypotheses regarding relationships among and between observed measures and their underlying latent constructs. A CFA model can be expressed in matrix equation form as

$$\mathbf{X} = \Lambda_x \xi + \delta,$$

where $\mathbf{X}$ is a $(NX \times 1)$ column vector of observed variables in deviation form; Λ_x (**Lambda X** or **LX**) is a $(NX \times NK)$ matrix of structural coefficients; ξ is a $(NK \times 1)$ column vector of latent variables; and δ is a $(NX \times 1)$ column vector of measurement error terms associated with the observed variables. In addition to the matrix Λ_x, a $(NK \times NK)$ variance/covariance matrix of latent variables, Φ, and a $(NX \times NX)$ variance/covariance matrix of measurement error terms, Θ_δ, must be specified to completely determine a particular CFA model.

Confirmatory factor analyses are used frequently in preparation for analyzing more general structural equation models that are introduced in the next chapter. However, in and of itself, CFA has become a powerful tool for the assessment of validity and reliability of, for example, academic achievement, personality, or behavior instruments. Traditionally, exploratory factor analysis (EFA) has been utilized to evaluate aspects of an instrument's construct validity [see the discussions in measurement books such as Allen and Yen (1979) and Crocker and Algina (1986)]. In an EFA, observed data—not the researcher's substantive knowledge and hypotheses—are used to explore and determine how many latent constructs underlie scales or items and how strongly the scales/items load on these data-determined factors. The investigator, in effect, pretends not to know (or, indeed, does not know?) anything about the probable factor structure of the instrument prior to data analysis. *If* "discovered" factors can be named appropriately and variables happen to load on these factors in an explicable fashion, EFA results can be interpreted as evidence of the measure's construct validity; otherwise, the analyst faces the problem of explaining and interpreting a more or less substantively meaningless factor structure.

Confirmatory factor analysis, on the other hand, is based on a philosophical approach to assessing psychometric properties that is fundamentally different from other variable reduction methods such as principal components or exploratory factor analysis. Substantive theory—rather than numerical data—is the driving force behind model conceptualization and evaluation. CFA should be viewed as a research *process* leading from model specification, identification, estimation, and assessment of data-model fit, to possible model modification and reestimation. During model specification, the analyst must hypothesize at the very least an initial structure among the observed variables, latent constructs, and measurement errors. The proposed model might be based on substantive knowledge and/or previous exploratory and theory-generating studies that were based on data sets other than the one currently under investigation. If the hypothesized structure remains very general, that is, if only a few a priori restrictions are placed on the model, its identification might be threatened and the estimation of some parameters from sample data might be impossible. For identified models and large enough representative samples, good parameter estimates usually are obtained by utilizing iterative methods such as maximum likelihood or generalized least squares. These methods—described in the next chapter—minimize the distance between the unrestricted and model-implied variance/covariance matrices. For overidentified models, the observed discrepancy between the two matrices is used as a basis for the assessment of overall data-model consistency.

Judgments regarding data-model fit or misfit in a confirmatory factor analysis should be based on several criteria. First, individual parameter estimates and associated statistics must be scrutinized for substantive and/or statistical impossibilities. Second, multiple overall fit indices should be considered since each was developed for a different purpose and comes with certain disadvantages (e.g., dependence on the multivariate normality assumption and sample size).

Finally, depending on the outcome of the fit assessment, a researcher might be justified to modify and reanalyze the initially proposed CFA model in order to improve data-model consistency. Sometimes, measures such as Lagrange Multipliers (modification indices) or Wald statistics can assist the analyst in deciding whether or not to free or fix a particular parameter. Again, model adjustments should be made only if they are theoretically and substantively justifiable. *A well-fitting model (initial or modified) by no means indicates that the true underlying structure has been found; data-model consistency in one particular investigation merely indicates that not enough evidence was present to reject the model*!

EXERCISES

2.1. Develop the six specific structural equations implied by the CFA model shown in Figures 2.6 and 2.7 by using its matrix representation in equation (2.37).

2.2. Use either the LISREL or EQS output file from the analysis of the *HBI* model in Figure 2.6 (Table 2.10 or 2.12, respectively) and equations (2.33) and (2.34) to compute the two parsimonious fit indices, PGFI and PNFI. What do you conclude?

2.3. Consider the data-model fit results in Table 2.3 from the LISREL analysis of the CFA model in Figure 2.1. Use the reported chi-square values—and their associated degrees of freedom—for the hypothesized and independence models to verify that

(a) NFI = .997,

(b) NNFI = .994, and

(c) CFI = .998,

by employing equations (2.29), (2.30), and (2.31), respectively.

2.4. Use the input of Table 2.1 to rerun the LISREL analysis of the model in Figure 2.1 and evaluate whether or not the model should have been modified. Base your decision on

(a) the modification indices and associated expected parameter change statistics; and

(b) the estimated expected value of the cross-validation index (ECVI) for the saturated, hypothesized, and independence models.

2.5. Reanalyze the CFA model in Figure 2.1 with the EQS package and verify the results by comparing the output to the one from the LISREL analysis partially shown in Tables 2.2 and 2.3.

2.6. (a) For the CFA model in Figure 2.5, specify the basic matrices Λ_x, Φ, and Θ_δ.

(b) Now reanalyze the model with (i) the LISREL command language and (ii) the SIMPLIS command language, and verify the results by comparing the output to the one from the EQS analysis partially shown in Table 2.5.

2.7. Modify the EQS input file shown in Table 2.4 to reflect the changes made to the corresponding CFA model after the identification of possible specification errors that were discussed in the text (see EQS Example 2.1). Verify your results by comparing them to the ones reproduced in Table 2.8.

2.8. Use either LISREL, SIMPLIS, or EQS and appropriate data from Appendix D to reanalyze the CFA models in Figures 2.1 and 2.5 for the female subsample.

(a) Compare the reliability estimates for the observed variables to those obtained from the analyses of data from the male subsample (for the model in Figure 2.1, use the R^2 values reported in Table 2.2; for the model in Figure 2.5, see Table 2.6).

(b) Compare data-model fit results to those obtained from the male subsample (for the model in Figure 2.1, see Table 2.3; for the model in Figure 2.5, consult Table 2.5).

(c) Do the analyses of the female subsample suggest different model modifications than those obtained from the analyses of the male subsample as discussed in the text?

2.9. Based on the available variables in Appendix D, conceptualize and theoretically justify an overidentified CFA model different from the ones presented in this chapter.

(a) Completely specify the model by expressing it in terms of the three basic matrices Λ_x, Φ, and Θ_δ.

(b) Ensure that the model is not underidentified—but possibly overidentified—by verifying that the number of free parameters, p, is less than the number of available (nonredundant) variances and covariances of observed variables, c.

(c) Use either LISREL, SIMPLIS, or EQS to estimate the proposed model.

(d) Assess the data-model fit by considering: (i) the plausibility of individual parameter estimates; (ii) the chi-square value compared to associated degrees of freedom; (iii) the GFI and AGFI; (iv) values of the NFI, NNFI, and CFI; and (v) the PGFI and PNFI.

(e) Substantively and statistically justify any modifications made to the initial model before reanalysis.

(f) Summarize and interpret the results.

2.10. Within your area of expertise and/or interest, identify a published research study that uses CFA for the data analysis and contains variance/covariance (or correlation and standard deviation) information for all observed variables.

(a) Use either LISREL, SIMPLIS, or EQS to reanalyze the data and verify the results published in the article.

(b) If not done already in the original study, use EQS to conduct Wald and Lagrange Multiplier tests and evaluate whether the results suggest possible modifications to the hypothesized model.

Recommended Readings

Two very useful, nontechnical treatments of some topics in confirmatory factor analysis that are not covered here (e.g., multigroup analyses) using the LISREL and EQS software packages are

Byrne, B.M. (1989). *A Primer of LISREL: Basic Applications and Programming for Confirmatory Factor Analytic Models.* New York: Springer-Verlag.

Byrne, B.M. (1994). *Structural Equation Modeling with EQS and EQS/Windows: Basic Concepts, Applications, and Programming.* Thousand Oaks, CA: Sage.

An excellent edited book, focusing on the data-model fit and model modification issues in SEM and including contributions by many experts, including Peter Bentler, Michael Browne, Robert Cudeck, Karl Jöreskog, and the late Jeff Tanaka, is

Bollen, K.A., and Long, J.S. (Eds.). (1993). *Testing Structural Equation Models.* Newbury Park, CA: Sage.

For a basic introduction to issues in confirmatory factor analysis, see

Long, J.S. (1983a). *Confirmatory Factor Analysis.* Beverly Hills: Sage.

Several books on social and behavioral science research methods include chapters on confirmatory factor analysis; for example, see

Nesselroade, J.R., and Cattell, R.B. (Eds.). (1988). *Handbook of Multivariate Experimental Psychology* (2nd ed.). New York: Plenum Press.

Pedhazur, E.J., and Schmelkin, L. (1991). *Measurement, Design, and Analysis: An Integrated Approach.* Hillsdale, NJ: Lawrence Erlbaum.

For a thorough treatment of validity and reliability and other issues in measurement from a classical true-score theory perspective, consult

Allen, M.J., and Yen, W.M. (1979). *Introduction to Measurement Theory*. Belmont, CA: Wadsworth.

Crocker, L., and Algina, J. (1986). *Introduction to Classical and Modern Test Theory*. Orlando, FL: Holt, Rinehart and Winston.

Finally, in-depth treatments of exploratory factor analysis can be found in

Gorsuch, R.L. (1983). *Factor Analysis* (2nd ed.). Hillsdale, NJ: Lawrence Erlbaum.

McDonald, R.P. (1985). *Factor Analysis and Related Methods*. Hillsdale, NJ: Lawrence Erlbaum.

Mulaik, S.A. (1972). *The Foundations of Factor Analysis*. New York: McGraw-Hill.

CHAPTER 3

General Structural Equation Modeling

Overview and Key Points

The most general structural equation models treated in this book are nothing more—and nothing less—than path analytical models (introduced in Chapter 1) that involve latent variables (discussed in Chapter 2). Even though classical path analysis has important advantages over conventional univariate or multivariate regression (e.g., the estimation of direct and indirect structural effects), one major disadvantage is that a priori hypothesized structures can be analyzed only under the usually unrealistic assumption that variables in the models are measured with no or negligible error. An integration of latent variables—as previously introduced in the context of confirmatory factor analysis—into path models relaxes this assumption and allows for the estimation of direct and indirect structural effects between variables or constructs that are not directly observable but, instead, are indicated by some imperfect observable measures.

Another disadvantage of classical path analysis as it was introduced in Chapter 1 is the dependency of the ordinary least squares (OLS) estimation method on several stringent assumptions, e.g., perfect measurement of observed variables and uncorrelated errors across equations. After a discussion of the specification and identification of structural equation models with latent variables, two estimation methods are presented (maximum likelihood and generalized least squares) that depend on less restrictive assumptions than OLS. Specifically, the following five key points are addressed in this chapter:

1. A general structural equation model is composed of three parts: the structural portion linking the latent variables and two measurement parts specifying how the observed exogenous/endogenous variables relate to the

latent exogenous/endogenous constructs. Specifically, a general structure is determined by the patterns of 0 and non-zero elements in eight basic matrices: (1) **Beta** (denoted as **B** or **BE**), a matrix of structural coefficients among latent endogenous variables; (2) **Gamma** (denoted as Γ or **GA**), a matrix of structural coefficients from latent exogenous to latent endogenous variables; (3) **Phi** (denoted as Φ or **PH**), a variance/covariance matrix of latent exogenous variables; (4) **Psi** (denoted as Ψ or **PS**), a variance/covariance matrix of error terms associated with latent endogenous variables; (5) **Lambda X** (denoted by Λ_X or **LX**), a matrix of structural coefficients linking the observed and latent exogenous variables; (6) **Theta Delta** (denoted by Θ_δ or **TD**), a variance/covariance matrix of measurement errors associated with the observed exogenous variables; (7) **Lambda Y** (denoted by Λ_Y or **LY**), a matrix of structural coefficients linking the observed and latent endogenous variables; and, finally, (8) **Theta Epsilon** (denoted by Θ_ε or **TE**), a variance/covariance matrix of measurement errors associated with the observed endogenous variables.

2. The path analytical and confirmatory factor analysis structures of the previous two chapters can be viewed as submodels of a general structural equation model.

3. The estimation of direct, indirect, and total structural effects among latent variables in a general structural equation model can be accomplished by using a matrix algorithm involving products of the structural coefficient matrices **Beta** and **Gamma**.

4. The methods of maximum likelihood (ML) and generalized least squares (GLS) allow for the consistent estimation of parameters in structural equation models involving latent variables but depend on the multivariate normality assumption. Both are techniques that iteratively minimize a specific fitting function of the unrestricted and model-implied variance/covariance matrix of observed variables.

5. Discussions and results from previous chapters regarding the issues of identification, data-model fit assessment, and model modification readily extend to general structural equation models.

As before, theoretical concepts are illustrated by annotated LISREL and EQS examples that are extensions of the structures introduced in previous chapters; corresponding SIMPLIS analyses can be found in Appendix A. Specifically, the examples involve data from (a) $n = 3094$ male respondents on the variables *MoEd*, *FaEd*, *PaJntInc*, *HSRank*, *AcAbilty*, *SelfConf*, *DegreAsp*, *Selctvty*, *Degree*, and *OcPrestg* (see Appendix D for coding information and summary statistics), and (b) $n = 167$ individuals that took the *Hutchins Behavior Inventory* (*HBI*) under two different specified situations and provided data on their sex and socioeconomic status (measured by *MoEd*, *FaEd*, and *FaOcc*; see Appendix E for coding information and summary statistics).

Specification and Identification of a General Structural Equation Model

The introductions of model specification and identification of path analytical and confirmatory factor analysis models in the previous two chapters should provide a good basis for understanding the specification and identification of more general structural equation models. In fact, grasping the topics discussed in the following sections mainly requires a conceptual integration of corresponding arguments in Chapters 1 and 2. To facilitate discussions, research questions associated with a hypothesized model previously introduced in Chapter 1 (see Figure 1.6 or 1.8 in LISREL or EQS Example 1.4) will serve as a guide; that is, the first set of illustrations will center around the estimation of direct and indirect structural effects among selected variables in an already familiar—but now more general—model of parents' and respondent's socioeconomic status.

Model Specification

Essentially, the specification of a general structural equation model involves three distinct tasks: first, a specific structure between latent exogenous and endogenous constructs must be hypothesized; second, it must be decided how to measure the exogenous latent variables; and, third, a measurement model for the endogenous latent constructs must be determined. To illustrate each of these steps, consider the path analytical model in Figure 3.1 that structurally relates two latent exogenous variables, ξ_s, parents' socioeconomic status ($\xi_1 = PaSES$) and academic rank ($\xi_2 = AcRank$), to the three latent endogenous constructs, η_r (eta), academic motivation ($\eta_1 = AcMotiv$), college prestige ($\eta_2 = ColgPres$), and respondent's socioeconomic status ($\eta_3 = SES$). Suppose the latent exogenous variables are measured by observed variables X_j (for $PaSES$: $X_1 = MoEd$, $X_2 = FaEd$, and $X_3 = PaJntInc$; for $AcRank$: $X_4 = HSRank$), and the latent endogenous variables are measured by observed variables Y_i (for $AcMotiv$: $Y_1 = AcAbilty$, $Y_2 = SelfConf$, and $Y_3 = DegreAsp$; for $ColgPres$: $Y_4 = Selctvty$; and for SES: $Y_5 = Degree$ and $Y_6 = OcPrestg$).

Note from Figure 3.1 that: (a) the overall structure among latent variables is the same as in the path analytical model in Figure 1.6, (b) the portion of the model involving the exogenous latent variables parents' SES and academic rank mirrors the confirmatory factor analysis model shown in Figure 2.1, and (c) the endogenous part of the model is similar to the CFA model in Figure 2.5, except that, here, a structure is specified among the latent variables academic motivation, college prestige, and respondent's SES.

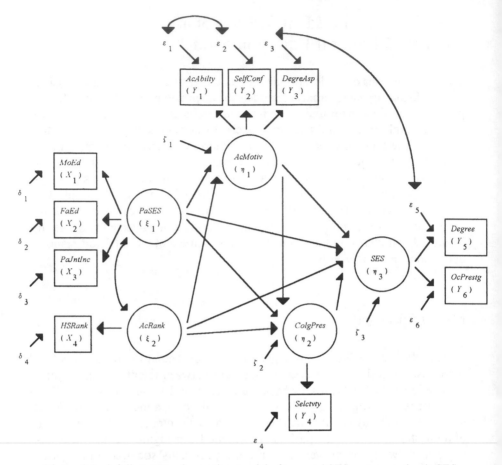

Figure 3.1. A full structural equation model of parents' SES on respondent's SES.

Now, first recall from Chapter 1 that a structural equation model involving only observed variables can be represented by the matrix equation

$$\mathbf{Y} = \mathbf{BY} + \mathbf{\Gamma X} + \zeta, \tag{3.1}$$

where $\mathbf{Y}$ is a $(NY \times 1)$ column vector of endogenous variables, $\mathbf{X}$ is a $(NX \times 1)$ column vector of exogenous variables measured as deviations from their means, $\mathbf{B}$ is a $(NY \times NY)$ matrix of structural coefficients from endogenous to other endogenous variables, $\mathbf{\Gamma}$ is a $(NY \times NX)$ matrix of structural coefficients from exogenous to endogenous variables, and ζ is a $(NY \times 1)$ column vector of error terms of endogenous variables.

In addition to the structural coefficient matrices $\mathbf{B}$ and $\mathbf{\Gamma}$, a $(NX \times NX)$ variance/covariance matrix $\mathbf{\Phi}$ of exogenous variables and a $(NY \times NY)$ variance/covariance matrix $\mathbf{\Psi}$ of elements in ζ must be specified to completely determine a path analytical model containing observed variables only (review Figure 1.3 for a schematic summary of the notation, order, and role of the four matrices $\mathbf{B}$, $\mathbf{\Gamma}$, $\mathbf{\Phi}$, and $\mathbf{\Psi}$).

Second, recall from Chapter 2 that a confirmatory factor analysis—or measurement—model relating the observed variables X_j, $j = 1, \ldots, NX$, to the latent *exogenous* constructs ξ_s, $s = 1, \ldots, NK$, can be represented by

$$\mathbf{X} = \mathbf{\Lambda}_X \mathbf{\xi} + \mathbf{\delta}, \tag{3.2}$$

where NX and NK denote the number of observed and latent exogenous variables, respectively; $\mathbf{X}$ is a $(NX \times 1)$ column vector of observed variables measured as deviations from their means; $\mathbf{\Lambda}_X$ is a $(NX \times NK)$ matrix of structural coefficients from latent exogenous to their observed indicator variables; $\mathbf{\xi}$ is a $(NK \times 1)$ column vector of latent exogenous constructs; and $\mathbf{\delta}$ is a $(NX \times 1)$ column vector of measurement error terms of the observed variables.

In addition to the structural coefficient matrix $\mathbf{\Lambda}_X$, a $(NK \times NK)$ variance/covariance matrix $\mathbf{\Phi}$ of latent exogenous variables and a $(NX \times NX)$ variance/covariance matrix $\mathbf{\Theta}_\delta$ of measurement error terms in $\mathbf{\delta}$ must be specified to completely determine a confirmatory factor analysis model (review Figure 2.3 for a summary of the notation, order, and role of the matrices $\mathbf{\Lambda}_X$, $\mathbf{\Phi}$, and $\mathbf{\Theta}_\delta$).

Similarly, it should be easy to see that the underlying structure relating the indicator variables Y_i, $i = 1, \ldots, NY$, to the latent *endogenous* constructs η_r, $r = 1, \ldots, NE$, can be expressed by a matrix equation of the form

$$\mathbf{Y} = \mathbf{\Lambda}_Y \mathbf{\eta} + \mathbf{\varepsilon}, \tag{3.3}$$

where NY and NE denote the number of observed and latent endogenous variables, respectively; $\mathbf{Y}$ is a $(NY \times 1)$ column vector of observed variables measured as deviations from their means; $\mathbf{\Lambda}_Y$ (**Lambda Y** or **LY**) is a $(NY \times NE)$ matrix of structural coefficients from latent endogenous variables to their observed indicators; $\mathbf{\eta}$ is a $(NE \times 1)$ column vector of latent endogenous constructs; and $\mathbf{\varepsilon}$ (**Epsilon** or **EP**) is a $(NY \times 1)$ column vector of measurement error terms of the observed variables.

Again, in addition to the structural coefficient matrix $\mathbf{\Lambda}_Y$, a $(NE \times NE)$ variance/covariance matrix $\mathbf{\Psi}$ of error terms associated with the latent variables and a $(NY \times NY)$ variance/covariance matrix $\mathbf{\Theta}_\varepsilon$ (**Theta Epsilon** or **TE**) of elements in $\mathbf{\varepsilon}$ must be specified to completely determine the measurement model relating the indicator variables Y to the latent endogenous variables η. Figure 3.2 summarizes the notation, order, and role of the three matrices $\mathbf{\Lambda}_Y$, $\mathbf{\Psi}$, and $\mathbf{\Theta}_\varepsilon$.

Finally, in order to represent a general structural equation model such as the one in Figure 3.1, which relates latent endogenous variables η_r to latent exogenous variables ξ_s, conceptually combine the structural equation model representation for observed variables [equation (3.1)] with the measurement model representations for the latent exogenous and endogenous variables [equations (3.2) and (3.3), respectively]. That is, a structural equation model

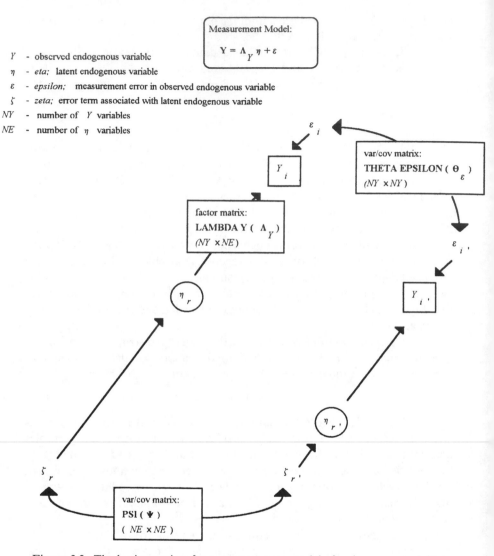

Figure 3.2. The basic matrices for a measurement model of endogenous variables.

involving latent constructs can be viewed as consisting of the following three components:

1. The structural portion relating the endogenous latent variables η_r, $r = 1$, ..., NE, to the exogenous latent variables ξ_s, $s = 1$, ..., NK, can be expressed by

$$\boldsymbol{\eta} = \mathbf{B}\boldsymbol{\eta} + \boldsymbol{\Gamma}\boldsymbol{\xi} + \boldsymbol{\zeta}, \tag{3.4}$$

where NE and NK denote the number of endogenous and exogenous latent variables, respectively; $\boldsymbol{\eta}$ is a $(NE \times 1)$ column vector of endogenous latent variables; $\boldsymbol{\xi}$ is a $(NK \times 1)$ column vector of exogenous latent variables; $\mathbf{B}$ is a $(NE \times NE)$ matrix of structural coefficients from endogenous to other endogenous latent variables; $\boldsymbol{\Gamma}$ is a $(NE \times NK)$ matrix of structural coefficients from exogenous to endogenous latent variables; and $\boldsymbol{\zeta}$ is a $(NE \times 1)$ column vector of error terms associated with the endogenous latent variables.

Specifically, the hypothesized structure among the five latent variables $(\xi_1 = PaSES, \xi_2 = AcRank, \eta_1 = AcMotiv, \eta_2 = ColgPres, \eta_3 = SES)$ in Figure 3.1 can be represented by equation (3.4) as

$$
\begin{bmatrix} \eta_1 \\ \eta_2 \\ \eta_3 \end{bmatrix} = \begin{bmatrix} 0 & 0 & 0 \\ \beta_{21} & 0 & 0 \\ \beta_{31} & \beta_{32} & 0 \end{bmatrix} \begin{bmatrix} \eta_1 \\ \eta_2 \\ \eta_3 \end{bmatrix} + \begin{bmatrix} \gamma_{11} & \gamma_{12} \\ \gamma_{21} & \gamma_{22} \\ \gamma_{31} & \gamma_{32} \end{bmatrix} \begin{bmatrix} \xi_1 \\ \xi_2 \end{bmatrix} + \begin{bmatrix} \zeta_1 \\ \zeta_2 \\ \zeta_3 \end{bmatrix}, \tag{3.5}
$$

which implies the following three structural equations:

$$ \eta_1 = \gamma_{11}\xi_1 + \gamma_{12}\xi_2 + \zeta_1, \tag{3.6} $$

$$ \eta_2 = \beta_{21}\eta_1 + \gamma_{21}\xi_1 + \gamma_{22}\xi_2 + \zeta_2, \tag{3.7} $$

and

$$ \eta_3 = \beta_{31}\eta_1 + \beta_{32}\eta_2 + \gamma_{31}\xi_1 + \gamma_{32}\xi_2 + \zeta_3. \tag{3.8} $$

Note that the set of equations (3.6)–(3.8) really is the same as the set of equations (1.37)–(1.39) of Chapter 1 with the only difference being that the observed exogenous and endogenous variables X and Y in Chapter 1 now are replaced with their corresponding latent counterparts, ξ and η.

2. The measurement model relating the exogenous observed variables X_j, $j = 1, \ldots, NX$, to the exogenous latent constructs ξ_s, $s = 1, \ldots, NK$ is given by

$$ \mathbf{X} = \boldsymbol{\Lambda}_X\boldsymbol{\xi} + \boldsymbol{\delta}, \tag{3.9} $$

where NX and NK denote the number of observed and latent exogenous variables, respectively, and the matrices $\mathbf{X}$, $\boldsymbol{\Lambda}_X$, $\boldsymbol{\xi}$, and $\boldsymbol{\delta}$ are defined as in equation (3.2). For the structural equation model of Figure 3.1, equation (3.9) becomes

$$
\begin{bmatrix} X_1 \\ X_2 \\ X_3 \\ X_4 \end{bmatrix} = \begin{bmatrix} \lambda_{11} & 0 \\ \lambda_{21} & 0 \\ \lambda_{31} & 0 \\ 0 & \lambda_{42} \end{bmatrix} \begin{bmatrix} \xi_1 \\ \xi_2 \end{bmatrix} + \begin{bmatrix} \delta_1 \\ \delta_2 \\ \delta_3 \\ \delta_4 \end{bmatrix}, \tag{3.10}
$$

which implies the same set of four structural equations that were discussed in Chapter 2 [see equations (2.1)–(2.4)].

3. The measurement model relating the endogenous observed variables Y_i, $i = 1, \ldots, NY$, to the exogenous latent constructs η_r, $r = 1, \ldots, NE$,

can be expressed by the equation

$$\mathbf{Y} = \boldsymbol{\Lambda}_Y \boldsymbol{\eta} + \boldsymbol{\varepsilon}, \tag{3.11}$$

where NY and NE denote the number of observed and latent endogenous variables, respectively, and the matrices $\mathbf{Y}$, $\boldsymbol{\Lambda}_Y$, $\boldsymbol{\eta}$, and $\boldsymbol{\varepsilon}$ are defined as in equation (3.3). For the model in Figure 3.1, equation (3.11) becomes

$$
\begin{bmatrix} Y_1 \\ Y_2 \\ Y_3 \\ Y_4 \\ Y_5 \\ Y_6 \end{bmatrix} =
\begin{bmatrix} \lambda_{11} & 0 & 0 \\ \lambda_{21} & 0 & 0 \\ \lambda_{31} & 0 & 0 \\ 0 & \lambda_{42} & 0 \\ 0 & 0 & \lambda_{53} \\ 0 & 0 & \lambda_{63} \end{bmatrix}
\begin{bmatrix} \eta_1 \\ \eta_2 \\ \eta_3 \end{bmatrix} +
\begin{bmatrix} \varepsilon_1 \\ \varepsilon_2 \\ \varepsilon_3 \\ \varepsilon_4 \\ \varepsilon_5 \\ \varepsilon_6 \end{bmatrix}, \tag{3.12}
$$

implying the following set of six structural equations that relate each of the endogenous observed variables Y_i to a latent construct η_r:

$$Y_1 = \lambda_{11}\eta_1 + \varepsilon_1, \tag{3.13}$$

$$Y_2 = \lambda_{21}\eta_1 + \varepsilon_2, \tag{3.14}$$

$$Y_3 = \lambda_{31}\eta_1 + \varepsilon_3, \tag{3.15}$$

$$Y_4 = \lambda_{42}\eta_2 + \varepsilon_4, \tag{3.16}$$

$$Y_5 = \lambda_{53}\eta_3 + \varepsilon_5, \tag{3.17}$$

and

$$Y_6 = \lambda_{63}\eta_3 + \varepsilon_6. \tag{3.18}$$

Paralleling discussions in Chapters 1 and 2, in addition to the four structural matrices $\mathbf{B}$, $\boldsymbol{\Gamma}$, $\boldsymbol{\Lambda}_X$, and $\boldsymbol{\Lambda}_Y$, a set of variance/covariance matrices needs to be specified to completely determine a general structural equation model. Specifically, the following four matrices are needed: (1) a $(NK \times NK)$ variance/covariance matrix $\boldsymbol{\Phi}$ of the latent exogenous variables; (2) a $(NE \times NE)$ variance/covariance matrix $\boldsymbol{\Psi}$ of error terms associated with the model-implied structural equations; (3) a $(NX \times NX)$ variance/covariance matrix $\boldsymbol{\Theta}_\delta$ of measurement errors of the observed exogenous variables; and (4) a $(NY \times NY)$ variance/covariance matrix $\boldsymbol{\Theta}_\varepsilon$ (**Theta Epsilon** or **TE**) of measurement error terms associated with the observed endogenous variables. For the model in Figure 3.1, the latter four matrices take the forms

$$\boldsymbol{\Phi} = \begin{bmatrix} \sigma^2_{\xi_1} & \\ \sigma_{\xi_2\xi_1} & \sigma^2_{\xi_2} \end{bmatrix}, \tag{3.19}$$

$$\boldsymbol{\Psi} = \begin{bmatrix} \sigma^2_{\zeta_1} & & \\ 0 & \sigma^2_{\zeta_2} & \\ 0 & 0 & \sigma^2_{\zeta_3} \end{bmatrix}, \tag{3.20}$$

$$\boldsymbol{\Theta}_\delta = \begin{bmatrix} \sigma^2_{\delta_1} & & & \\ 0 & \sigma^2_{\delta_2} & & \\ 0 & 0 & \sigma^2_{\delta_3} & \\ 0 & 0 & 0 & \sigma^2_{\delta_4} \end{bmatrix}, \tag{3.21}$$

and

$$\boldsymbol{\Theta}_\varepsilon = \begin{bmatrix} \sigma^2_{\varepsilon_1} & & & & & \\ \sigma_{\varepsilon_2\varepsilon_1} & \sigma^2_{\varepsilon_2} & & & & \\ 0 & 0 & \sigma^2_{\varepsilon_3} & & & \\ 0 & 0 & 0 & \sigma^2_{\varepsilon_4} & & \\ 0 & 0 & \sigma_{\varepsilon_5\varepsilon_3} & 0 & \sigma^2_{\varepsilon_5} & \\ 0 & 0 & 0 & 0 & 0 & \sigma^2_{\varepsilon_6} \end{bmatrix}. \tag{3.22}$$

Note the hypothesized non-zero elements in the $\boldsymbol{\Theta}_\varepsilon$ matrix, $\sigma_{\varepsilon_2\varepsilon_1}$ and $\sigma_{\varepsilon_5\varepsilon_3}$, corresponding to the proposed non-zero covariances between the measurement error terms of the variable pairs *SelfConf* and *AcAbilty* and *Degree* and *DegreAsp* in Figure 3.1. These hypothesized covariances are a result of a modification to the confirmatory factor analysis model of Figure 2.5, as was discussed in EQS Example 2.1. Also, recall from Chapter 1 that correlated error terms violate an assumption of OLS parameter estimation; later in this chapter, estimation techniques are introduced that are suitable for estimating such associations.

In conclusion then, the patterns of 0 and non-zero elements in the eight matrices $\mathbf{B}$, $\boldsymbol{\Gamma}$, $\boldsymbol{\Phi}$, $\boldsymbol{\Psi}$, $\boldsymbol{\Lambda}_X$, $\boldsymbol{\Theta}_\delta$, $\boldsymbol{\Lambda}_Y$, and $\boldsymbol{\Theta}_\varepsilon$ completely determine a general structural equation model. Figure 3.3—a combination of Figures 1.3, 2.3, and 3.2—summarizes the notation, order, and role of these eight basic matrices. With $\boldsymbol{\Lambda}_X = \mathbf{I}$, $\boldsymbol{\Theta}_\delta = \mathbf{0}$, $\boldsymbol{\Lambda}_Y = \mathbf{I}$, and $\boldsymbol{\Theta}_\varepsilon = \mathbf{0}$, note that the general structure reduces to a representation of the classical path analysis models of Chapter 1; and, similarly, with $\mathbf{B} = \mathbf{0}$, $\boldsymbol{\Gamma} = \mathbf{0}$, $\boldsymbol{\Psi} = \mathbf{0}$, $\boldsymbol{\Lambda}_Y = \mathbf{0}$, and $\boldsymbol{\Theta}_\varepsilon = \mathbf{0}$, the general model becomes a confirmatory factor analysis as presented in Chapter 2. That is, the classical path analysis and CFA modes of previous chapters can be viewed as submodels of the general structures introduced here.

Combining the assumptions stated in Chapters 1 and 2 for path analytical and confirmatory factor analysis models, we must accept the following seven statistical assumptions for general structural equation models:

1. The exogenous and endogenous latent variables have a mean of 0 $[E(\xi) = E(\eta) = 0]$.
2. The structural relations from the exogenous to the endogenous latent variables are linear.
3. The error terms in ζ in equation (3.4): (a) have a mean of 0 $[E(\zeta) = 0]$ and a constant variance across observations; (b) are independent, i.e., uncorrelated across observations; and (c) are uncorrelated with the exogenous latent variables $[E(\xi\zeta') = E(\zeta\xi') = 0]$.
4. The matrix $(\mathbf{I} - \mathbf{B})$ is nonsingular, i.e., the matrix is invertible.

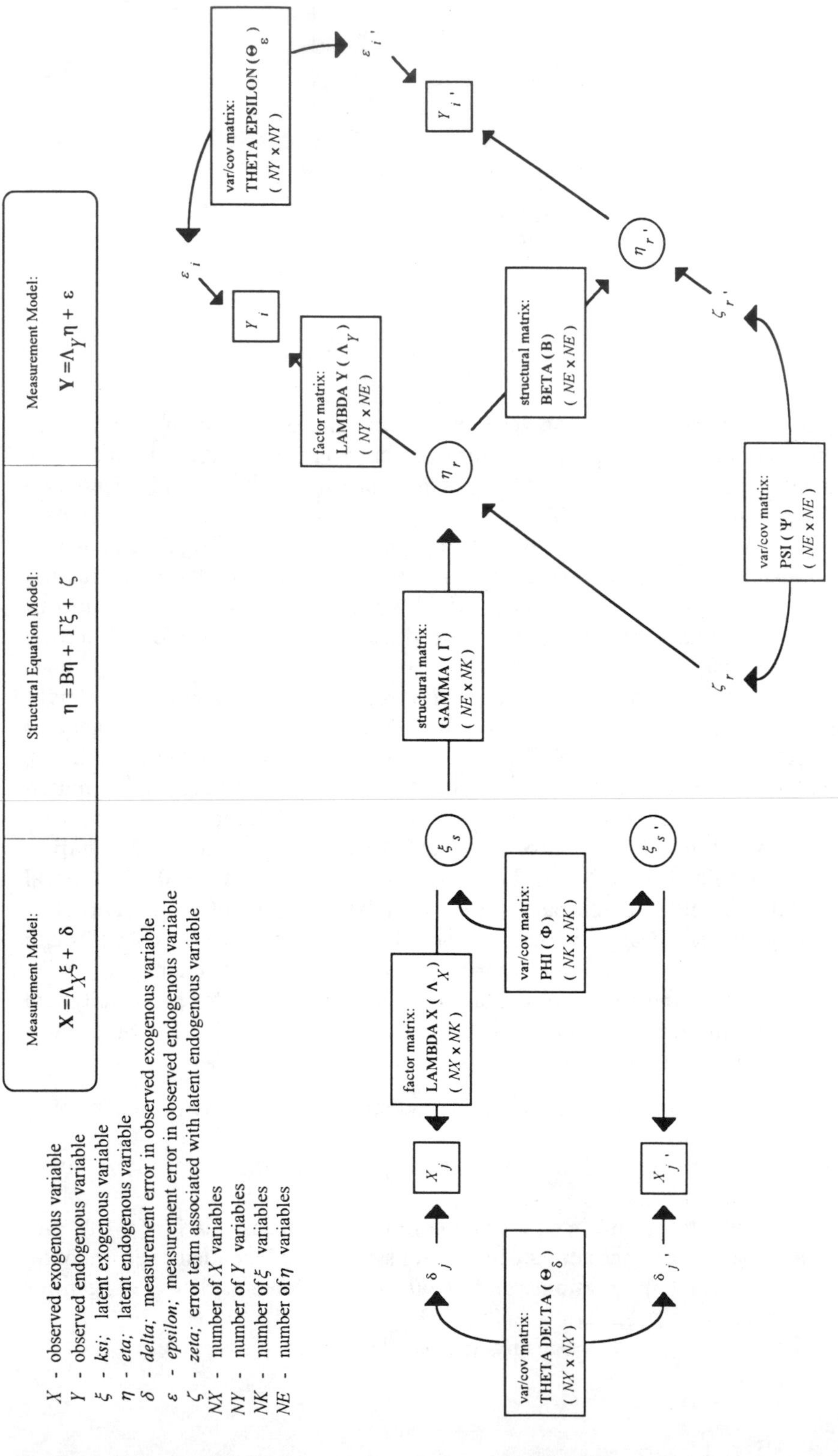

Figure 3.3.   The basic matrices for general structural equation models.

5. The means of the exogenous and endogenous observed variables are 0, i.e., $E(\mathbf{X}) = E(\mathbf{Y}) = \mathbf{0}$.
6. The relationships between the (exogenous and endogenous) indicator variables and their (exogenous and endogenous) associated latent constructs are linear.
7. The measurement error terms in $\boldsymbol{\delta}$ and $\boldsymbol{\varepsilon}$ in equations (3.9) and (3.11): (a) have a mean of 0 $[E(\boldsymbol{\delta}) = E(\boldsymbol{\varepsilon}) = \mathbf{0}]$ and a constant variance across observations; (b) are independent, i.e., elements in $\boldsymbol{\delta}$ are uncorrelated across observations, and the same holds for elements in $\boldsymbol{\varepsilon}$; (c) are uncorrelated with the exogenous and endogenous latent variables $[E(\boldsymbol{\xi}\boldsymbol{\delta}') = E(\boldsymbol{\delta}\boldsymbol{\xi}') = \mathbf{0};$ $E(\boldsymbol{\eta}\boldsymbol{\delta}') = E(\boldsymbol{\delta}\boldsymbol{\eta}') = \mathbf{0}; \ E(\boldsymbol{\eta}\boldsymbol{\varepsilon}') = E(\boldsymbol{\varepsilon}\boldsymbol{\eta}') = \mathbf{0};$ and $E(\boldsymbol{\xi}\boldsymbol{\varepsilon}') = E(\boldsymbol{\varepsilon}\boldsymbol{\xi}') = \mathbf{0}]$; and (d) are uncorrelated with each other $[E(\boldsymbol{\varepsilon}\boldsymbol{\delta}') = E(\boldsymbol{\delta}\boldsymbol{\varepsilon}') = \mathbf{0}]$.

As was mentioned in Chapter 2, assumptions 1 and 5 here are not necessary if an analysis of mean structures is being considered. Furthermore, assumption 7(d) can be relaxed by specifying non-zero elements in a matrix $\boldsymbol{\Theta}_{\delta\varepsilon}$ (**Theta Delta-Epsilon**) of the covariances between elements in $\boldsymbol{\delta}$ and $\boldsymbol{\varepsilon}$. Since the specification of $\boldsymbol{\Theta}_{\delta\varepsilon}$ is not necessary in most straightforward applications, the reader is referred to the LISREL manual for further details and illustration (also, see Exercise 3.1).

Identification

As with the path analytical and confirmatory factor analysis models of previous chapters, the identification status of general structural equation models (i.e., under-, just-, or overidentified) is rather difficult to prove mathematically. For general recursive models, however, it usually suffices to ensure that:

1. The number of free parameters p in the eight basic matrices ($\mathbf{B}, \boldsymbol{\Gamma}, \boldsymbol{\Phi}, \boldsymbol{\Psi}, \boldsymbol{\Lambda}_X, \boldsymbol{\Theta}_\delta, \boldsymbol{\Lambda}_Y,$ and $\boldsymbol{\Theta}_\varepsilon$) does not exceed the number of available (nonredundant) variances and covariances among the $(NX + NY)$ observed variables c, where now $c = (NX + NY)(NX + NY + 1)/2$.
2. All exogenous and endogenous latent variables (ξ_s and η_r) have an assigned unit of measurement (see Chapter 2).
3. An observed exogenous or endogenous variable that is the only indicator of a latent (exogenous or endogenous) variable is assumed to be measured without error.

For the model in Figure 3.1, the first rule is satisfied since there are currently $p = 37$ free parameters [for verification, count the number of free parameters in the eight basic matrices; see equations (3.5), (3.10), (3.12), and (3.19)–(3.22)] with $c = (4 + 6)(4 + 6 + 1)/2 = 55$ nonredundant variances and covariances. Recall from Chapters 1 and 2, however, that the condition $p \leq c$ is necessary *but not sufficient* for model identification. As specified, the model in Figure 3.1 is not identified. To rectify this problem, modify the appropriate model-implied matrices so that the remaining two rules are satisfied. First, the assignment of units of measurement to the latent variables in the model can be accomplished by specifying reference variables. For example, the matrices Λ_X and Λ_Y in equations (3.10) and (3.12), respectively, could be changed to

$$\Lambda_X = \begin{bmatrix} 1 & 0 \\ \lambda_{21} & 0 \\ \lambda_{31} & 0 \\ 0 & 1 \end{bmatrix} \tag{3.23}$$

and

$$\Lambda_Y = \begin{bmatrix} 1 & 0 & 0 \\ \lambda_{21} & 0 & 0 \\ \lambda_{31} & 0 & 0 \\ 0 & 1 & 0 \\ 0 & 0 & 1 \\ 0 & 0 & \lambda_{63} \end{bmatrix}. \tag{3.24}$$

The two fixed elements in Λ_X in equation (3.23) ($\lambda_{11} = 1$ and $\lambda_{42} = 1$) imply units of measurement for the latent exogenous variables *PaSES* and *AcRank* equal to those of the observed variables *MoEd* and *HSRank*, respectively. Similarly, the fixed elements in Λ_Y in equation (3.24) ($\lambda_{11} = 1$, $\lambda_{42} = 1$, and $\lambda_{53} = 1$) assign units of measurement to the latent endogenous constructs *AcMotiv*, *ColgPres*, and *SES* equal to those of the indicators *AcAbilty*, *Selctvty*, and *Degree*, respectively.

Second, note from Figure 3.1 that there are two latent variables that have one indicator only: The construct academic rank is measured by *HSRank* (X_4) while college prestige is measured by *Selctvty* (Y_4). Thus, the two observed variables are assumed to be measured without error, which changes the variance/covariance matrices Θ_δ and Θ_ε in equations (3.21) and (3.22) to

$$\Theta_\delta = \begin{bmatrix} \sigma_{\delta_1}^2 & & & \\ 0 & \sigma_{\delta_2}^2 & & \\ 0 & 0 & \sigma_{\delta_3}^2 & \\ 0 & 0 & 0 & 0 \end{bmatrix} \tag{3.25}$$

and

$$\Theta_\varepsilon = \begin{bmatrix} \sigma_{\varepsilon_1}^2 & & & & & \\ \sigma_{\varepsilon_2\varepsilon_1} & \sigma_{\varepsilon_2}^2 & & & & \\ 0 & 0 & \sigma_{\varepsilon_3}^2 & & & \\ 0 & 0 & 0 & 0 & & \\ 0 & 0 & \sigma_{\varepsilon_5\varepsilon_3} & 0 & \sigma_{\varepsilon_5}^2 & \\ 0 & 0 & 0 & 0 & 0 & \sigma_{\varepsilon_6}^2 \end{bmatrix}. \tag{3.26}$$

With the above changes in the matrices Λ_X, Λ_Y, Θ_δ, and Θ_ε, the model in Figure 3.1 becomes overidentified with $df = c - p = 55 - 30 = 25$ degrees of freedom. Before providing an illustration of the estimation of a general structural equation model by conducting a LISREL analysis of the model of parents' on respondent's socioeconomic status, the concepts of direct, indirect, and total structural effects that were introduced in Chapter 1 are briefly reviewed and then redefined in more general terms.

The Direct, Indirect, and Total Structural Effect Components

In Chapter 1, the estimation of direct and indirect effects among any two variables in a path model was shown to be one of the advantages of classical path analysis over multiple regression. Recall that: (a) the direct effect (DE) of an endogenous on another endogenous observed variable was defined as the coefficient in the $\mathbf{B}$ matrix that is associated with the two variables; (b) the direct effect of an exogenous on an endogenous observed variable is the structural coefficient in the $\boldsymbol{\Gamma}$ matrix that is associated with the two variables; (c) a particular indirect effect between two observed variables through specific intervening variables is computed by forming the product of structural coefficients in the $\mathbf{B}$ and/or $\boldsymbol{\Gamma}$ matrices along the particular path from the independent to the dependent variable; and (d) the total indirect effect (IE) between two observed variables is defined as the sum of *all* particular indirect effects through possible intervening variables. Finally, the total effect (TE) between two observed variables was defined as the sum of the direct and total indirect effect components.

The definitions of direct, indirect, and total effects remain unchanged for structural equation models that involve latent variables. However, rather than using the First Law of Path Analysis [equation (1.27)], a more convenient method to estimate the various effect components is available and is used by SEM programs such as LISREL and EQS. For an illustration, first consider the direct, indirect, and total effects of the latent endogenous construct $AcMotiv$ (η_1) on SES (η_3) in the model of Figure 3.1. Clearly, all direct effects of latent endogenous on other endogenous factors ($DE_{\eta\eta}$) are contained in the structural coefficient matrix $\mathbf{B}$ [see equation (3.5)]; specifically, $DE_{\eta_2\eta_1} = \beta_{31}$. Now, for a convenient estimation of the indirect effect through an intervening variable, consider the matrix product $\mathbf{B}^2 = \mathbf{BB}$, that is,

$$\mathbf{B}^2 = \begin{bmatrix} 0 & 0 & 0 \\ \beta_{21} & 0 & 0 \\ \beta_{31} & \beta_{32} & 0 \end{bmatrix} \begin{bmatrix} 0 & 0 & 0 \\ \beta_{21} & 0 & 0 \\ \beta_{31} & \beta_{32} & 0 \end{bmatrix} = \begin{bmatrix} 0 & 0 & 0 \\ 0 & 0 & 0 \\ \beta_{32}\beta_{21} & 0 & 0 \end{bmatrix}. \qquad (3.27)$$

This matrix contains in its third row and first column the particular indirect effect between η_3 (SES) and η_1 ($AcMotiv$) through the intervening latent construct η_2 (*ColgPres*). It is the only non-zero element in $\mathbf{B}^2$, indicating that there are no other indirect effects between the latent endogenous variables. More generally, for any structural equation model, the matrix $\mathbf{B}^2$ contains all specific indirect effects between endogenous constructs that involve the multiplication of two structural coefficients (i.e., indirect effects through *one* intervening variable). Similarly, the matrix $\mathbf{B}^3$ contains all particular indirect effects between endogenous variables that involve *three* structural coefficients, that is, via *two* intervening variables. Continuing this logic, it follows that the total indirect effects ($IE_{\eta\eta}$; the sum of all particular indirect effects) between any two endogenous variables η are given as elements in the infinite matrix series

$$IE_{\eta\eta} = \mathbf{B}^2 + \mathbf{B}^3 + \mathbf{B}^4 + \cdots. \tag{3.28}$$

In the above example, $\mathbf{B}^3$ and all subsequent powers of $\mathbf{B}$ are equal to the $\mathbf{0}$ matrix since, for the model in Figure 3.1, there are no indirect effects between endogenous latent variables through more than one intervening construct. Generalizing this observation, note that, for *recursive* structural equation models involving NE latent endogenous variables, the matrix $\mathbf{B}^{NE}$ (and all subsequent powers of $\mathbf{B}$) always will be equal to the $\mathbf{0}$ matrix, guaranteeing that the series in equation (3.28) converges.

Finally, since a total effect (TE) between any two variables in a model is defined as the sum of the direct and total indirect effects, it follows that total effects between latent endogenous variables η are given as elements in the matrix

$$TE_{\eta\eta} = DE_{\eta\eta} + IE_{\eta\eta} = \mathbf{B} + \mathbf{B}^2 + \mathbf{B}^3 + \mathbf{B}^4 + \cdots. \tag{3.29}$$

Thus, for the model in Figure 3.1,

$$TE_{\eta\eta} = \mathbf{B} + \mathbf{B}^2 = \begin{bmatrix} 0 & 0 & 0 \\ \beta_{21} & 0 & 0 \\ \beta_{31} & \beta_{32} & 0 \end{bmatrix} + \begin{bmatrix} 0 & 0 & 0 \\ 0 & 0 & 0 \\ \beta_{32}\beta_{21} & 0 & 0 \end{bmatrix}$$

$$= \begin{bmatrix} 0 & 0 & 0 \\ \beta_{21} & 0 & 0 \\ \beta_{31} + \beta_{32}\beta_{21} & \beta_{32} & 0 \end{bmatrix}. \tag{3.30}$$

In particular, the total structural effect of the construct *AcMotiv* (η_1) on *SES* (η_3) is given in equation (3.30) as $TE_{\eta_3\eta_1} = \beta_{31} + \beta_{32}\beta_{21}$, which can be verified easily by consulting Figure 3.1 and following the procedures outlined in Chapter 1.

Next, consider the direct, indirect, and total effects between latent exogenous and endogenous variables; specifically, focus on the effects of *AcRank* (ξ_2) on *SES* (η_3). Again, all direct effects from exogenous to endogenous variables ($DE_{\eta\xi}$) are given in a structural coefficient matrix—this time, in the matrix $\boldsymbol{\Gamma}$ [see equation (3.5)]; that is, for example, $DE_{\eta_3\xi_2} = \gamma_{32}$. Since all particular indirect effects between exogenous and endogenous variables involve products of coefficients from the $\mathbf{B}$ and $\boldsymbol{\Gamma}$ matrices, consider the infinite matrix series

$$IE_{\eta\xi} = \mathbf{B}\boldsymbol{\Gamma} + \mathbf{B}^2\boldsymbol{\Gamma} + \mathbf{B}^3\boldsymbol{\Gamma} + \mathbf{B}^4\boldsymbol{\Gamma} + \cdots = (\mathbf{B} + \mathbf{B}^2 + \mathbf{B}^3 + \mathbf{B}^4 + \cdots)\boldsymbol{\Gamma}. \quad (3.31)$$

From the above discussion we know that, for the model in Figure 3.1, $\mathbf{B}^3$ and all subsequent powers of $\mathbf{B}$ are equal to the $\mathbf{0}$ matrix so that, here, equation (3.31) becomes

$$
\begin{aligned}
IE_{\eta\xi} &= \mathbf{B}\boldsymbol{\Gamma} + \mathbf{B}^2\boldsymbol{\Gamma} = (\mathbf{B} + \mathbf{B}^2)\boldsymbol{\Gamma} \\
&= \begin{bmatrix} 0 & 0 & 0 \\ \beta_{21} & 0 & 0 \\ \beta_{31} + \beta_{32}\beta_{21} & \beta_{32} & 0 \end{bmatrix} \begin{bmatrix} \gamma_{11} & \gamma_{12} \\ \gamma_{21} & \gamma_{22} \\ \gamma_{31} & \gamma_{32} \end{bmatrix} \\
&= \begin{bmatrix} 0 & 0 \\ \beta_{21}\gamma_{11} & \beta_{21}\gamma_{12} \\ (\beta_{31} + \beta_{32}\beta_{21})\gamma_{11} + \beta_{32}\gamma_{21} & (\beta_{31} + \beta_{32}\beta_{21})\gamma_{12} + \beta_{32}\gamma_{22} \end{bmatrix} \\
&= \begin{bmatrix} 0 & 0 \\ \beta_{21}\gamma_{11} & \beta_{21}\gamma_{12} \\ \beta_{31}\gamma_{11} + \beta_{32}\beta_{21}\gamma_{11} + \beta_{32}\gamma_{21} & \beta_{31}\gamma_{12} + \beta_{32}\beta_{21}\gamma_{12} + \beta_{32}\gamma_{22} \end{bmatrix}.
\end{aligned}
\tag{3.32}
$$

The element in the third row and second column of this matrix is the total indirect effect of *AcRank* (ξ_2) on *SES* (η_3); that is, $IE_{\eta_3\xi_2} = \beta_{31}\gamma_{12} + \beta_{32}\beta_{21}\gamma_{12} + \beta_{32}\gamma_{22}$, as was shown previously in Chapter 1; see equation (1.43). Similarly, the total indirect effect of *PaSES* (ξ_1) on *SES* is given in equation (3.32) by $IE_{\eta_3\xi_1} = \beta_{31}\gamma_{11} + \beta_{32}\beta_{21}\gamma_{11} + \beta_{32}\gamma_{21}$, which can be verified by consulting Figure 3.1 and following the procedures discussed in Chapter 1.

Finally, the total effects between exogenous and endogenous variables are given in a matrix $TE_{\eta\xi}$, where

$$
\begin{aligned}
TE_{\eta\xi} = DE_{\eta\xi} + IE_{\eta\xi} &= \boldsymbol{\Gamma} + \mathbf{B}\boldsymbol{\Gamma} + \mathbf{B}^2\boldsymbol{\Gamma} + \mathbf{B}^3\boldsymbol{\Gamma} + \mathbf{B}^4\boldsymbol{\Gamma} + \cdots \\
&= (\mathbf{I} + \mathbf{B} + \mathbf{B}^2 + \mathbf{B}^3 + \mathbf{B}^4 + \cdots)\boldsymbol{\Gamma}
\end{aligned}
\tag{3.33}
$$

since total effects are defined as the sum of direct and total indirect effect components. Thus, for the model in Figure 3.1,

$$
\begin{aligned}
TE_{\eta\xi} = DE_{\eta\xi} + IE_{\eta\xi} &= \begin{bmatrix} \gamma_{11} & \gamma_{12} \\ \gamma_{21} & \gamma_{22} \\ \gamma_{31} & \gamma_{32} \end{bmatrix} \\
&\quad + \begin{bmatrix} 0 & 0 \\ \beta_{21}\gamma_{11} & \beta_{21}\gamma_{12} \\ \beta_{31}\gamma_{11} + \beta_{32}\beta_{21}\gamma_{11} + \beta_{32}\gamma_{21} & \beta_{31}\gamma_{12} + \beta_{32}\beta_{21}\gamma_{12} + \beta_{32}\gamma_{22} \end{bmatrix} \\
&= \begin{bmatrix} \gamma_{11} & \gamma_{12} \\ \gamma_{21} + \beta_{21}\gamma_{11} & \gamma_{22} + \beta_{21}\gamma_{12} \\ \gamma_{31} + \beta_{31}\gamma_{11} + \beta_{32}\beta_{21}\gamma_{11} + \beta_{32}\gamma_{21} & \gamma_{32} + \beta_{31}\gamma_{12} + \beta_{32}\beta_{21}\gamma_{12} + \beta_{32}\gamma_{22} \end{bmatrix}.
\end{aligned}
\tag{3.34}
$$

Table 3.1. Direct, Indirect, and Total Effect Components for General Structural Equation Models

Effect Component	Exogenous (ξ) → Endogenous (η)	Endogenous (η) → Endogenous (η)
direct (*DE*)	$\mathbf{\Gamma}$	$\mathbf{B}$
indirect (*IE*)	$(\mathbf{I} - \mathbf{B})^{-1}\mathbf{\Gamma} - \mathbf{\Gamma}$	$(\mathbf{I} - \mathbf{B})^{-1} - \mathbf{I} - \mathbf{B}$
total (*TE*)	$(\mathbf{I} - \mathbf{B})^{-1}\mathbf{\Gamma}$	$(\mathbf{I} - \mathbf{B})^{-1} - \mathbf{I}$

In conclusion, for computational convenience and accuracy it can be shown mathematically (see, for example, Hayduk, 1987, Chapter 8) that the presented matrix expression of indirect and total effects among variables in a general structural equation model [equations (3.28), (3.29), (3.31), and (3.33)] are equivalent to the corresponding entries in Table 3.1. Software packages such as LISREL and EQS use these expressions to compute the various effect components in path analytical and more general models [note that direct, indirect, and total effects of *latent* (ξ and/or η) on *observed* variables (*Y*) can be computed; since the interpretation of these occurs relatively infrequently in the literature, they are not discussed here; but see, for example, Hayduk (1987) or the LISREL manual].

LISREL EXAMPLE 3.1: THE DIRECT AND INDIRECT STRUCTURAL EFFECTS IN A MODEL OF PARENTS' SES AND RESPONDENT'S ACADEMIC RANK AND SES. The first illustration of an analysis of a structural equation model involving latent variables is a LISREL analysis of the model in Figure 3.1. In Chapter 1, several research questions regarding the relationship between parents' and respondent's socioeconomic status were raised [first, from a simple and multiple linear regression point of view (see LISREL and EQS Examples 1.1 and 1.2) and, second, within the context of the path analytical models in Figures 1.1 and 1.6 (see LISREL and EQS Examples 1.3 and 1.4)]. Here, essentially the same questions are addressed but now incorporating concepts and results from a priori confirmatory factor analyses of the exogenous and endogenous portions of the full model (see Chapter 2; LISREL and EQS Examples 2.1). Extensive discussions of the interpretation of obtained results (e.g., the interpretation of the direct, indirect, and total effect components) are omitted here since, conceptually, they would not differ from those in Chapter 1. That is, *while not explicitly repeated, all cautionary notes regarding the interpretation of SEM results mentioned in previous chapters* (see, for example, discussions in LISREL Example 1.4) *are equally applicable to the general structural equation models discussed here.*

LISREL Input. A possible LISREL program file for the analysis of the model in Figure 3.1 is shown in Table 3.2. The correlation matrix and standard deviation from all 15 observed variables listed in Appendix D are used as input data (see lines 6 through 23). Thus, variables in Figure 3.1 must be

Table 3.2. LISREL Input File for the Model in Figure 3.1

1	Example 3.1. A Structural Equation Model of Parents' on Respondent's SES
2	DA NI = 15 NO = 3094
3	LA
4	MoEd FaEd PaJntInc HSRank FinSucc ConCollg AcAbilty DriveAch SelfConf
5	DegreAsp ColContr Selctvty Degree OcPrestg Income
6	KM SY
7	1
8	.610 1
9	.446 .531 1
10	.115 .128 .055 1
11	−.077 −.097 −.016 −.052 1
12	−.203 −.216 −.393 .002 −.018 1
13	.192 .216 .154 .493 −.086 −.079 1
14	−.042 −.017 −.023 .205 .063 .010 .251 1
15	.090 .112 .068 .269 −.021 −.043 .487 .327 1
16	.116 .122 .101 .194 −.008 .021 .236 .195 .206 1
17	.139 .205 .170 .049 −.125 .011 .119 .018 .056 .106 1
18	.255 .300 .293 .372 −.111 −.114 .382 .152 .216 .214 .294 1
19	.117 .129 .141 .189 −.025 −.067 .242 .184 .179 .253 .144 .254 1
20	.057 .084 .059 .153 −.002 .017 .163 .098 .090 .125 .110 .155 .481 1
21	.012 −.008 .093 .037 .157 −.060 .064 .096 .040 .025 −.020 .074 .106 .136 1
22	SD
23	1.229 1.511 2.649 .777 .847 .612 .744 .801 .782 1.014 .475 1.990 .962 1.591 1.627
24	SE
25	AcAbilty SelfConf DegreAsp Selctvty Degree OcPrestg MoEd FaEd PaJntInc HSRank/
26	MO NX = 4 NK = 2 NY = 6 NE = 3 c
27	BE = SD GA = FU PH = SY PS = DI c
28	LX = FU,FI TD = SY,FI c
29	LY = FU,FI TE = SY,FI
30	LK
31	PaSES AcRank
32	LE
33	AcMotiv ColgPres SES
34	VA 1.0 LX(1,1) LX(4,2)
35	FR LX(2,1) LX(3,1)
36	FR TD(1,1) TD(2,2) TD(3,3)
37	VA 1.0 LY(1,1) LY(4,2) LY(5,3)
38	FR LY(2,1) LY(3,1) LY(6,3)
39	FR TE(1,1) TE(2,2) TE(3,3) TE(5,5) TE(6,6) TE(2,1) TE(5,3)
40	OU EF ND = 3

SElected by using the SE command (lines 24 and 25), which must end with a slash (/) since not all input variables are selected for the analysis.

The MOdel section on lines 26 through 29 includes specifications of the number of observed and latent variables involved in the model [NX = 4 (*MoEd, FaEd, PaJntInc,* and *HSRank*); NK = 2 (*PaSES* and *AcRank*); NY = 6 (*AcAbilty, SelfConf, DegreAsp, Selctvty, Degree,* and *OcPrestg*); and NE = 3 (*AcMotiv, ColgPres,* and *SES*)] before giving initial settings for the eight basic matrices **B**, **Γ**, **Φ**, **Ψ**, Λ_X, Θ_δ, Λ_Y, and Θ_ε. Note that the settings and subsequent specifications of the free or fixed status of parameters associated with the structural portion of the model and the measurement model of the *exogenous* variables (lines 27, 28, and 34 through 36) mirror the previous specifications in Tables 1.13 and 2.1. Finally, the program setup for the measurement portion of the *endogenous* variables (lines 29 and 37 through 39) is similar to the one associated with the exogenous variables with the exception that—in addition to appropriate diagonal elements—two off-diagonal elements in the matrix Θ_ε, TE(2,1) and TE(5,3), are specified as free parameters (see line 39 and Figure 3.1; also, consider Exercise 2.6).

Selected LISREL Output. A partial output from the above analysis is shown in Table 3.3 and includes the unstandardized maximum likelihood estimates of elements in the eight matrices and the unstandardized total and indirect effects among latent variables in the model (see Exercise 3.1). First, consider the portions of the output that deal with the measurement portions of the model (i.e., coefficients in the matrices Λ_X, Θ_δ, Λ_Y, and Θ_ε). As expected, estimates and associated test statistics of the factor loadings and variances and covariances of measurement error terms are very similar to those obtained from the confirmatory factor analyses in Chapter 2 (see LISREL and EQS Example 2.1, Tables 2.2 and 2.8). Also, note that the reliability estimates for the exogenous and endogenous indicator variables—the coefficients of determination R^2 associated with each observed variable—from the present analysis are close to those obtained previously from separate confirmatory factor analyses. For example, an analysis of the model in Figure 3.1 resulted in a reliability estimate of 0.705 for the variable *FaEd*, while the CFA of the model in Figure 2.1 produced an estimate of 0.729. Similarly, Table 3.3 shows a reliability of 0.115 for the endogenous indicator *DegreAsp*, and it is easy to calculate from the standardized results in Table 2.8 that the corresponding estimate is 0.147 from an analysis of the modified CFA model in Figure 2.5.

Second, consider the portion of the output that relates to the structural portion of the model in Figure 3.1, that is, the estimated structural coefficients (the direct effects among latent constructs in the model) in the matrices **B** and **Γ**. Here, the results seem not to mirror those obtained from the previous path analysis of a model of the same structure that involved observed variables only and no non-zero covariances between measurement errors (see Figure 1.6 and Table 1.14). Specifically, while three estimates in the

Table 3.3. Partial LISREL Output from an Analysis of the Model in
Figure 3.1.

LISREL ESTIMATES (MAXIMUM LIKELIHOOD)

LAMBDA-Y

	AcMotiv	ColgPres	SES
AcAbilty	1.000	—	—
SelfConf	0.612 (0.029) 20.777	—	—
DegreAsp	0.641 (0.044) 14.421	—	—
Selctvty	—	1.000	—
Degree	—	—	1.000
OcPrestg	—	—	1.101 (0.086) 12.769

LAMBDA-X

	PaSES	AcRank
MoEd	1.000	—
FaEd	1.434 (0.044) 32.838	—
PaJntInc	1.904 (0.063) 30.459	—
HSRank	—	1.000

BETA

	AcMotiv	ColgPres	SES
AcMotiv	—	—	—
ColgPres	1.564 (0.239) 6.545	—	—
SES	0.620 (0.115) 5.404	0.039 (0.012) 3.106	—

GAMMA

	PaSES	AcRank
AcMotiv	0.165 (0.014) 11.403	0.436 (0.015) 29.480

Table 3.3 (*cont.*)

ColgPres	0.478	0.151
	(0.058)	(0.111)
	8.224	1.366
SES	0.016	−0.080
	(0.026)	(0.048)
	0.599	−1.638

PHI

	PaSES	AcRank
PaSES	0.782	
	(0.039)	
	20.087	
AcRank	0.099	0.604
	(0.014)	(0.015)
	7.032	39.326

PSI

AcMotiv	ColgPres	SES
0.138	2.659	0.540
(0.019)	(0.092)	(0.051)
7.126	28.956	10.569

SQUARED MULTIPLE CORRELATIONS FOR STRUCTURAL EQUATIONS

AcMotiv	ColgPres	SES
0.520	0.329	0.183

THETA-EPS

	AcAbilty	SelfConf	DegreAsp	Selctvty	Degree	OcPrestg
AcAbilty	0.265					
	(0.020)					
	13.261					
SelfConf	0.107	0.504				
	(0.014)	(0.016)				
	7.564	31.866				
DegreAsp	—	—	0.910			
			(0.024)			
			37.308			
Selctvty	—	—	—	—		
Degree	—	—	0.110	—	0.259	
			(0.015)		(0.050)	
			7.321		5.216	
OcPrestg	—	—	—	—	—	1.730
						(0.074)
						23.246

Table 3.3 (*cont.*)

SQUARED MULTIPLE CORRELATIONS FOR Y-VARIABLES

AcAbilty	SelfConf	DegreAsp	Selctvty	Degree	OcPrestg
0.521	0.177	0.115	1.000	0.718	0.316

THETA-DELTA

MoEd	FaEd	PaJntInc	HSRank
0.729	0.674	4.182	—
(0.027)	(0.043)	(0.130)	
27.034	15.724	32.275	

SQUARED MULTIPLE CORRELATIONS FOR X-VARIABLES

MoEd	FaEd	PaJntInc	HSRank
0.518	0.705	0.404	1.000

TOTAL AND INDIRECT EFFECTS
TOTAL EFFECTS OF KSI ON ETA

	PaSES	AcRank
AcMotiv	0.165	0.436
	(0.014)	(0.015)
	11.403	29.480
ColgPres	0.736	0.833
	(0.043)	(0.041)
	17.199	20.163
SES	0.146	0.222
	(0.021)	(0.022)
	6.932	10.319

INDIRECT EFFECTS OF KSI ON ETA

	PaSES	AcRank
AcMotiv	—	—
ColgPres	0.258	0.681
	(0.044)	(0.105)
	5.806	6.495
SES	0.131	0.302
	(0.018)	(0.044)
	7.436	6.798

TOTAL EFFECTS OF ETA ON ETA

	AcMotiv	ColgPres	SES
AcMotiv	—	—	—
ColgPres	1.564	—	—
	(0.239)		
	6.545		

Table 3.3 (*cont.*)

SES	0.680	0.039	—
	(0.109)	(0.012)	
	6.222	3.106	

INDIRECT EFFECTS OF ETA ON ETA

	AcMotiv	ColgSelc	SES
AcMotiv	—	—	—
ColgSelc	—	—	—
SES	0.060	—	—
	(0.017)		
	3.536		

Γ matrix in Table 3.3 are not statistically different from 0 (the direct effects of *PaSES* on *SES*, *AcRank* on *ColgPres*, and *AcRank* on *SES*), the results in Table 1.14 indicate the statistical significance of corresponding effects among observed variables (*FaEd* on *Degree*, *HSRank* on *Selctvty*, and *HSRank* on *Degree*) in the model in Figure 1.6. Furthermore, the coefficients of determination associated with the three model-implied structural equations indicate that: (a) 52% of the variance in the construct academic motivation can be accounted for by variability in the exogenous factors parents' SES and academic rank; (b) about 33% of the variance in college prestige is explained by the three constructs parents' SES, academic rank, and academic motivation; and (c) approximately 18% of the variability in the latent variable respondent's SES is accounted for by the other four factors in the model of Figure 3.1 (i.e., *PaSES*, *AcRank*, *AcMotiv*, and *ColgPres*). Especially note that, while the exogenous variables in the path analytical model in Figure 1.6 (*FaEd* and *HSRank*) accounted for only about 5% of the variance in the indicator variable *DegreeAsp* of the construct *AcMotiv* ($R^2 = 0.047$; see Table 1.14), here, the corresponding coefficient of determination increased more than tenfold.

Third, and finally, all estimates of the indirect and total effects among the five latent variables at the bottom of Table 3.3 are statistically significant. It is evident that the total structural effect of parents' on respondent's socioeconomic status (0.146, $t = 6.932$), for example, is due mainly to a significant indirect effect via the intervening constructs academic motivation and college prestige (0.131, $t = 7.436$) rather than a nonsignificant direct influence ($\gamma_{31} = 0.016$, $t = 0.599$). Similarly, respondent's academic rank seems to influence the prestige of the college attended only indirectly through academic motivation since its total structural effect (0.833, $t = 20.163$) is composed of a significant indirect influence (0.681, $t = 6.495$) but a nonsignificant direct effect ($\gamma_{32} = 0.151$, $t = 1.366$).

Parameter Estimation

In most structural equation modeling situations, estimation techniques such as maximum likelihood (ML) or generalized least squares (GLS) are preferred over the ordinary least squares method (OLS; see Chapter 1) since they allow for the analysis of models involving latent variables and non-zero error covariances across structural equations. After defining the model-implied variance/covariance matrix for general structural equation models, the ML and GLS techniques are discussed and, since the former has been used throughout the book, only the latter will be illustrated by an EQS analysis of the model in Figure 3.1.

The Model-Implied Variance/Covariance Matrix

The distinction between the unrestricted and model-implied variance/covariance matrix was made in the context of the confirmatory factor analysis models of Chapter 2 [see equation (2.9)]. For the more general models discussed here, obtaining an expression of the latter matrix involves writing the former matrix as a function of the eight basic matrices $\mathbf{B}, \mathbf{\Gamma}, \mathbf{\Phi}, \mathbf{\Psi}$, $\Lambda_X, \Theta_\delta, \Lambda_Y$, and Θ_ε. Specifically, consider the unrestricted variance/covariance matrix Σ of the observed exogenous and endogenous variables X_j and Y_i in a general structural equation model,

$$\Sigma = E\left(\begin{bmatrix} \mathbf{Y} \\ \mathbf{X} \end{bmatrix}\begin{bmatrix} \mathbf{Y} \\ \mathbf{X} \end{bmatrix}'\right), \tag{3.35}$$

where

$$\begin{bmatrix} \mathbf{Y} \\ \mathbf{X} \end{bmatrix}' = [Y_1 \ Y_2 \ \cdots \ Y_{NY} \ X_1 \ X_2 \ \cdots \ X_{NX}]$$

is the $[1 \times (NY + NX)]$ row vector containing the $(NY + NX)$ observed variables in a particular structural equation model. Using rules outlined in Appendix C and substituting the right sides of equations (3.9) and (3.11) for the vectors $\mathbf{X}$ and $\mathbf{Y}$, respectively, yields

$$\Sigma = E\left(\begin{bmatrix} \mathbf{Y} \\ \mathbf{X} \end{bmatrix}\begin{bmatrix} \mathbf{Y} \\ \mathbf{X} \end{bmatrix}'\right) = E\left(\begin{bmatrix} \mathbf{YY'} & \mathbf{YX'} \\ \mathbf{XY'} & \mathbf{XX'} \end{bmatrix}\right)$$

$$= E\left(\begin{bmatrix} (\Lambda_Y\eta + \varepsilon)(\Lambda_Y\eta + \varepsilon)' & (\Lambda_Y\eta + \varepsilon)(\Lambda_X\xi + \delta)' \\ (\Lambda_X\xi + \delta)(\Lambda_Y\eta + \varepsilon)' & (\Lambda_X\xi + \delta)(\Lambda_X\xi + \delta)' \end{bmatrix}\right). \tag{3.36}$$

Expanding the last expression of Σ in equation (3.36), taking expected values, and using the definitions of—and assumptions involving—the eight basic matrices, it can be shown that the model-implied variance/covariance matrix

$\Sigma(\theta)$ for a general structural equation model involving latent variables can be expressed as (see Exercise 3.3)

$$\Sigma(\theta) = \begin{bmatrix} \Lambda_Y(I - B)^{-1}(\Gamma\Phi\Gamma' + \Psi)[(I - B)^{-1}]'\Lambda_Y' + \Theta_\varepsilon & \Lambda_Y(I - B)^{-1}\Gamma\Phi\Lambda_X' \\ \Lambda_X\Phi\Gamma'[(I - B)^{-1}]'\Lambda_Y' & \Lambda_X\Phi\Lambda_X' + \Theta_\delta \end{bmatrix}.$$
(3.37)

From equation (3.37) it is easy to infer the model-implied variance/covariance matrix for the path analytical models of Chapter 1 since any classical path model can be viewed as a submodel of the general structure with $\Lambda_X = I$, $\Theta_\delta = 0$, $\Lambda_Y = I$, and $\Theta_\varepsilon = 0$. Substituting the right sides of these identities into equation (3.37) yields the following expression of $\Sigma(\theta)$ for structural equation models involving observed variables only (i.e., the path analysis models discussed in Chapter 1):

$$\Sigma(\theta) = \begin{bmatrix} (I - B)^{-1}(\Gamma\Phi\Gamma' + \Psi)[(I - B)^{-1}]' & (I - B)^{-1}\Gamma\Phi \\ \Phi\Gamma'[(I - B)^{-1}]' & \Phi \end{bmatrix}.$$
(3.38)

Similarly, the confirmatory factor analysis models of Chapter 2 are submodels of the general structure with $B = 0$, $\Gamma = 0$, $\Psi = 0$, $\Lambda_Y = 0$, and $\Theta_\varepsilon = 0$. Substitution of these values into equation (3.37) reduces the matrix to its lower right-hand quadrant, which equals the expression of $\Sigma(\theta)$ for CFA models that was previously given in equation (2.9).

Iterative Methods

The overall task in iteratively estimating parameters in an identified structural equation model is to find matrix estimates $\hat{B}$, $\hat{\Gamma}$, $\hat{\Phi}$, $\hat{\Psi}$, $\hat{\Lambda}_X$, $\hat{\Theta}_\delta$, $\hat{\Lambda}_Y$, and $\hat{\Theta}_\varepsilon$ of the eight basic matrices so that the model-implied variance/covariance matrix $\Sigma(\hat{\theta})$ is as "close" as possible to the unrestricted variance/covariance matrix Σ (recall from Chapter 2 that $\hat{\theta}$ is a vector containing estimates of the p model-implied parameters). Suppose that initial estimates $\tilde{B}$, $\tilde{\Gamma}$, $\tilde{\Phi}$, $\tilde{\Psi}$, $\tilde{\Lambda}_X$, $\tilde{\Theta}_\delta$, $\tilde{\Lambda}_Y$, and $\tilde{\Theta}_\varepsilon$ are available and $\Sigma(\tilde{\theta})$ is judged to be close enough to Σ. Then these estimates are taken to be the "best" and final, and the iteration process is terminated. If, however, $\Sigma(\tilde{\theta})$ does not reproduce Σ to an acceptable degree, new and improved estimates, say $\tilde{\tilde{B}}$, $\tilde{\tilde{\Gamma}}$, $\tilde{\tilde{\Phi}}$, $\tilde{\tilde{\Psi}}$, $\tilde{\tilde{\Lambda}}_X$, $\tilde{\tilde{\Theta}}_\delta$, $\tilde{\tilde{\Lambda}}_Y$, and $\tilde{\tilde{\Theta}}_\varepsilon$, of the eight matrices need to be computed so that $\Sigma(\tilde{\tilde{\theta}})$ is closer to Σ than $\Sigma(\tilde{\theta})$. The process of improving parameter estimates based on those obtained during the previous iteration is repeated until satisfactory closeness of the model-implied and unrestricted variance/covariance matrix is achieved.

To measure closeness, a suitable *fitting function*, $F[S, \Sigma(\theta)]$, of the two variance/covariance matrices S (the sample estimate of Σ) and $\Sigma(\theta)$ is utilized that satisfies the general conditions that: (a) the values of $F[S, \Sigma(\theta)]$ are nonnegative scalars, and (b) $F[S, \Sigma(\theta)] = 0$ if and only if $S = \Sigma(\theta)$. That is, the fitting function is iteratively minimized with $F[S, \Sigma(\tilde{\tilde{\theta}})] < F[S, \Sigma(\tilde{\theta})]$, implying that the vector $\tilde{\tilde{\theta}}$ contains "better" parameter estimates of elements in

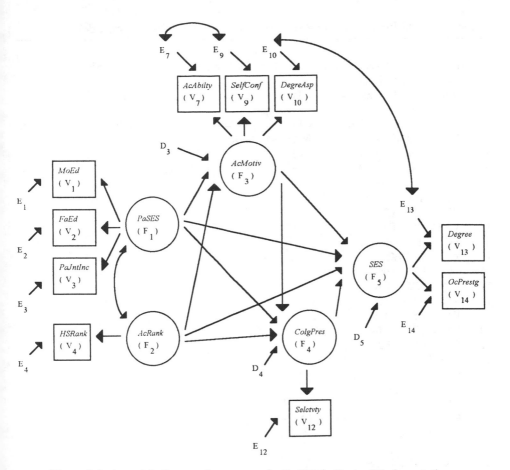

Figure 3.4. A model of parents' on respondent's SES in Benter-Weeks notation.

the basic matrices than $\tilde{\theta}$. The final estimates—collected in the vector $\hat{\theta}$—correspond to the minimum value of the fitting function, $F[\mathbf{S}, \mathbf{\Sigma}(\hat{\theta})]$, that is used for judging the fit of the data to the hypothesized model, as was discussed in Chapter 2.

The methods of maximum likelihood (ML) and generalized least squares (GLS) are similar in approach and in their underlying assumptions (in fact, they are asymptotically equivalent; consult the LISREL or EQS manuals for further details) and differ only in the fitting function that is utilized during the iteration process. For a correctly specified model and a sufficiently large data set drawn from a multivariate normal distribution, both methods provide consistent parameter estimates, appropriate standard errors and test statistics, and $(n - 1)$ times the minimum value of the fitting function is distributed as an approximate χ^2-statistic with $df = (c - p)$, where c denotes the number of nonredundant elements in $\mathbf{\Sigma}$ and p is the number of free parameters in the

Table 3.4. EQS Input File for a GLS Analysis of the Model in Figure 3.4

1	/TITLE
2	Example 3.1. A GLS Analysis of a Model of Parents' on Respondent's SES
3	/SPECIFICATIONS
4	Variables = 15; Cases = 3094; Matrix = Correlation; Method = GLS;
5	/LABELS
6	V1 = MoEd; V2 = FaEd; V3 = PaJntInc; V4 = HSRank; V5 = FinSucc; V6 = ConCollg;
7	V7 = AcAbilty; V8 = DriveAch; V9 = SelfConf; V10 = DegreAsp; V11 = ColContr;
8	V12 = Selctvty; V13 = Degree; V14 = OcPrestg; V15 = Income;
9	F1 = PaSES; F2 = AcRank; F3 = AcMotiv; F4 = ColgPres; F5 = SES;
10	/EQUATIONS
11	V1 = F1 + E1;
12	V2 = *F1 + E2;
13	V3 = *F1 + E3;
14	V4 = F2 + E4;
15	V7 = F3 + E7;
16	V9 = *F3 + E9;
17	V10 = *F3 + E10;
18	V12 = F4 + E12;
19	V13 = F5 + E13;
20	V14 = *F5 + E14;
21	F3 = *F1 + *F2 + D3;
22	F4 = *F1 + *F2 + *F3 + D4;
23	F5 = *F1 + *F2 + *F3 + *F4 + D5;
24	/VARIANCES
25	F1, F2 = *;
26	E1, E2, E3, E7, E9, E10, E13, E14 = *;
27	E4, E12 = 0;
28	D3 to D5 = *;
29	/COVARIANCES
30	F2, F1 = *;
31	E7, E9 = *;
32	E10, E13 = *;
33	D5, D3 = 0;
34	D5, D4 = 0;
35	D4, D3 = 0;
36	/MATRIX
37	1
38	.610 1
39	.446 .531 1
40	.115 .128 .055 1
41	−.077 −.097 −.016 −.052 1
42	−.203 −.216 −.393 .002 −.018 1
43	.192 .216 .154 .493 −.086 −.079 1
44	−.042 −.017 −.023 .205 .063 .010 .251 1
45	.090 .112 .068 .269 −.021 −.043 .487 .327 1
46	.116 .122 .101 .194 −.008 .021 .236 .195 .206 1
47	.139 .205 .170 .049 −.125 .011 .119 .018 .056 .106 1

Table 3.4 (*cont.*)

48	.255 .300 .293 .372 $-$.111 $-$.114 .382 .152 .216 .214 .294 1
49	.117 .129 .141 .189 $-$.025 $-$.067 .242 .184 .179 .253 .144 .254 1
50	.057 .084 .059 .153 $-$.002 .017 .163 .098 .090 .125 .110 .155 .481 1
51	.012 $-$.008 .093 .037 .157 $-$.060 .064 .096 .040 .025 $-$.020 .074 .106 .136 1
52	/STANDARD DEVIATIONS
53	1.229 1.511 2.649 .777 .847 .612 .744 .801 .782 1.014 .475 1.990 .962 1.591 1.627
54	/PRINT
55	Effects = Yes;
56	/END

vector θ. The ML and GLS fitting functions can be expressed as

$$F_{ML}[S, \Sigma(\theta)] = \ln|\Sigma(\theta)| - \ln|S| + tr[S\Sigma(\theta)^{-1}] - (NX + NY) \quad (3.39)$$

and

$$F_{GLS}[S, \Sigma(\theta)] = \tfrac{1}{2} tr([S^{-1}(S - \Sigma(\theta))]^2). \quad (3.40)$$

Note from equations (3.39) and (3.40) that both functions are nonnegative and become 0 if and only if $S = \Sigma(\theta)$ since the trace of a matrix (tr) simply is the sum of its diagonal elements.

Finally, it should be mentioned that a third estimation technique, the Asymptotically Distribution Free method (ADF; Browne, 1984), does not depend on the underlying distribution of the data and, thus, is the most general method of those mentioned. However, one major disadvantage of the ADF is that it yields incorrect chi-square statistics for small samples. In LISREL, Browne's ADF is available under the name Generally Weighted Least Squares (WLS) while EQS refers to the ADF method as Arbitrary distribution Generalized Least Squares (AGLS). Here, the ADF method is not discussed further since a demonstration would require either the use of the program PRELIS (Jöreskog and Sörbom, 1993c) for a LISREL illustration or the analysis of raw data for an EQS run.

EQS EXAMPLE 3.1: A GENERALIZED LEAST SQUARES ANALYSIS OF A MODEL OF PARENTS' SES AND RESPONDENT'S ACADEMIC RANK AND SES. For an illustration of a GLS analysis of a general structural equation model with the EQS package, again consider the model of parents' on respondent's SES (Figure 3.1) that is reproduced in Figure 3.4 in the Benter-Weeks notation. In the figure, the observed variables (V) are numbered according to their order in the input matrix (see Table 3.4). Also, the error terms—or Disturbances— associated with the structural equations (that is, the terms ζ_r, $r = 1, 2, \ldots$, NE, in the Jöreskog-Keesling-Wiley notation) are denoted by the letters D according to Bentler-Weeks.

Table 3.5. Selected EQS Output from the GLS Analysis of the Model in Figure 3.4

GENERALIZED LEAST SQUARES SOLUTION
(NORMAL DISTRIBUTION THEORY)

MEASUREMENT EQUATIONS WITH STANDARD ERRORS AND
TEST STATISTICS

MOED = V1 =   1.000 F1 +   1.000 E1

FAED = V2 =   1.440*F1 +   1.000 E2
                      .044
                      32.671

PAJNTINC = V3 =   1.916*F1 +   1.000 E3
                      .063
                      30.193

HSRANK = V4 =   1.000 F2 +   1.000 E4

ACABILTY = V7 =   1.000 F3 +   1.000 E7

SELFCONF = V9 =   .615*F3 +   1.000 E9
                      .029
                      20.873

DEGREASP = V10 =   .650*F3 +   1.000 E10
                      .045
                      14.545

SELCTVTY = V12 =   1.000 F4 +   1.000 E12

DEGREE = V13 =   1.000 F5 +   1.000 E13

OCPRESTG = V14 =   1.104*F5 +   1.000 E14
                      .086
                      12.843

CONSTRUCT EQUATIONS WITH STANDARD ERRORS AND
TEST STATISTICS

ACMOTIV = F3 =   .165*F1 +   438*F2 +   1.000 D3
                      .014   .015
                      11.407   29.463

COLGPRES = F4 =   1.547*F3 +   474*F1 +   .191*F2 +   1.000 D4
                      .234   .057   .110
                      6.602   8.250   1.743

SES = F5 =   .628*F3 +   .032*F4 +   .018*F1 +   −.069*F2 +   1.000 D5
                      .114   .013   .026   .048
                      5.490   2.559   .687   −1.454

Table 3.5 (*cont.*)

VARIANCES OF INDEPENDENT VARIABLES

V		F		
	I	F1- PASES	.775*	I
	I		.039	I
	I		19.952	I
	I			I
	I	F2 -ACRANK	.598*	I
	I		.015	I
	I		39.058	I
	I			I

VARIANCES OF INDEPENDENT VARIABLES

E			D		
E1 -MOED	.726*	I	D3 -ACMOTIV	.137*	I
	.027	I		.019	I
	26.666	I		7.114	I
		I			I
E2 -FAED	.670*	I	D4 -COLGPRES	2.589*	I
	.043	I		.089	I
	15.462	I		29.047	I
		I			I
E3 -PAJNTINC	4.019*	I	D5 -SES	.537*	I
	.128	I		.051	I
	31.520	I		10.602	I
		I			I
E7 -ACABILTY	.266*	I			I
	.020	I			I
	13.407	I			I
		I			I
E9 -SELFCONF	.495*	I			I
	.015	I			I
	32.088	I			I
		I			I
E10 -DEGREASP	.883*	I			I
	.024	I			I
	36.645	I			I
		I			I
E13 -DEGREE	.253*	I			I
	.049	I			I
	5.142	I			I
		I			I
E14 -OCPRESTG	1.711*	I			I
	.074	I			I
	23.075	I			I
		I			I

Table 3.5 (*cont.*)

COVARIANCES AMONG INDEPENDENT VARIABLES

V		F		
	I F2 -ACRANK	.099*	I
	I F1 -PASES	.014	I
	I	7.099	I
	I		I

COVARIANCES AMONG INDEPENDENT VARIABLES

E		D	
E9 -SELFCONF	.108* I		I
E7 -ACABILTY	.014 I		I
	7.943 I		I
	I		I
E13 -DEGREE	.097* I		I
E10 -DEGREASP	.015 I		I
	6.486 I		I
	I		I

EQS Input. The EQS input file for an analysis of the model in Figure 3.4 is shown in Table 3.4. The statement "Method = GLS;" in the /SPECIFICA-TIONS section on line 4 requests that the parameter estimates be based on minimizing the generalized least squares fitting function in equation (3.40) [if this statement is omitted, EQS uses the ML function in equation (3.39) by default].

The first ten relationships in the /EQUATIONS section in lines 11 through 20 specify how the observed variables relate to the five underlying constructs: Lines 11 through 14 correspond to specifying the structure of the measurement model of the exogenous variables [see equation (3.9)] while lines 15 through 20 correspond to the measurement model of the endogenous variables [see equation (3.11)]. The remaining three equations in lines 21 through 23 specify the hypothesized structure among the latent factors [see equation (3.4)]. Finally, the statements in the /VARIANCES and /COVARIANCES sections (lines 24 through 28 and lines 29 through 35, respectively) indicate which variances and covariances of independent variables (in the sense of Bentler-Weeks) are fixed to 0 and which are free to be estimated by the program; these specifications correspond to the forms of the model-implied matrices Φ, Ψ, Θ_δ, and Θ_ε in Jöreskog-Keesling-Wiley notation.

Selected EQS Output. Table 3.5 lists the generalized least squares estimates, standard errors, and test statistics of all free parameters in the model of

Figure 3.4. Estimates of the indirect and total effects among latent variables in the model are omitted from the table; but see Exercise 3.4. It suffices to compare the present results with the maximum likelihood estimates in Table 3.3 obtained from the LISREL analysis of the same model. Since the sample size is large ($n = 3094$) and GLS and ML are asymptotically equivalent, it should not be surprising to observe very similar results from the two estimation methods.

The Structural Equation Modeling Process: An Illustrated Review and Summary

In the Preface of this book structural equation modeling (SEM) was introduced as being a research *process*. The presentation of some of the statistical details in the preceding pages and chapters, however, might have lead to a perception of SEM as a mere statistical technique, losing track of the point that the topics discussed are not discreet ones but all part of a larger whole. The remaining sections of this chapter are devoted to illustrating the SEM process—from model conceptualization and specification, identification, estimation, and assessment of data-model fit, to possible model modification. No new concepts are introduced; rather, topics and issues previously introduced in detail are reviewed and illustrated by a LISREL and EQS analysis of a general structural equation model involving data from the $n = 167$ residence-hall counselors on the *Hutchins Behavior Inventory* (*HBI*; see Chapter 2, Example 2.2).

Model Specification and Identification

The *HBI* was introduced in Chapter 2 as a measure of an individual's thinking-feeling-acting (TFA) orientation; that is, does a person approach a particular situation or context with a primarily cognitive, affective, or psychomotor orientation (see Hutchins and Mueller, 1992)? Two possible research questions that could be of interest to counselors are (1) Do TFA orientations—as measured by the *HBI*—depend on the biological sex and the socioeconomic status of the respondents? and (2) Is the *HBI* indeed sensitive to the situational dependence of individuals' TFA orientations, as was theoretically hypothesized by Hutchins (1979, 1982)?

Consider the structural equation model in Figure 3.5 that shows the hypothesized relationships between respondent's biological sex (*BioSex* $= \xi_1$), socioeconomic status (*SES* $= \xi_2$), situation specificity (*Situatin* $= \xi_3$), and the behavior constructs *Thinking* (η_1), *Feeling* (η_2), and *Acting* (η_3). With regard to the structural portion of the model (i.e., the relationships among the latent

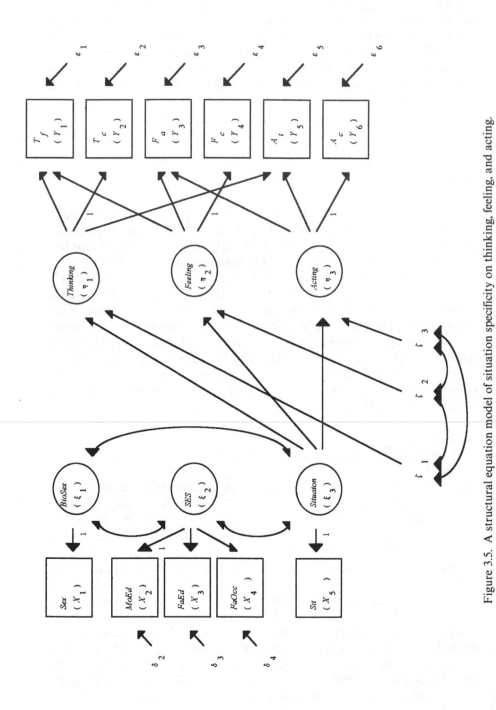

Figure 3.5. A structural equation model of situation specificity on thinking, feeling, and acting.

variables ξ_1, ξ_2, ξ_3, η_1, η_2, and η_3), note that only the variable Situatin is hypothesized to influence the constructs *Thinking*, *Feeling*, and *Acting*; that is, TFA orientations are hypothesized to be independent of an individual's biological sex and socioeconomic status.

The measurement model of the exogenous variables shows what variables are used to measure the three latent exogenous constructs: (1) The observed variable *Sex* (X_1) serves as a single indicator of biological sex; (2) three variables—mother's educational level ($MoEd = X_2$), father's education ($FaEd = X_3$), and father's occupation ($FaOcc = X_4$)—measure respondent's socioeconomic status; and (3) the dichotomous variable situation ($Sit = X_5$—half of the $n = 167$ residence-hall counselors responded to the *HBI* under the situation "How do I view myself as a student?" while the remaining half answered to the question "How do I view myself when confronted with a close friend in emotional distress?") is the sole indicator for situation specificity (see Appendix E for coding information on the observed variables).

Finally, the measurement model of the endogenous constructs *Thinking*, *Feeling*, and *Acting* essentially mirrors the CFA model of Figure 2.6 with the exception that, here, the *HBI* characteristic scales T_c, F_c, and A_c are used as reference variables to define the units of measurement for the latent constructs. Furthermore, since the latent factors are dependent variables in the current model (Figure 3.5), hypothesized associations among the three behavior constructs in the CFA model (Figure 2.6) are replaced by associations among corresponding error terms ζ_r.

For the purpose of model identification, note from Figure 3.5 that: (a) all latent variables have a unit of measurement equal to that of the assigned reference variables (indicated by the 1's next to the appropriate arrows in the figure) and (b) measurement errors of single indicator variables are specified to be 0 (indicated by the absence of appropriate error terms from the figure). The forms of the eight basic matrices that are implied by Figure 3.5 are as follows:

1. For the structural portion relating the latent constructs *BioSex*, *SES*, *Situatin*, *Thinking*, *Feeling*, and *Acting*,

$$\mathbf{B} = \begin{bmatrix} 0 & 0 & 0 \\ 0 & 0 & 0 \\ 0 & 0 & 0 \end{bmatrix}, \tag{3.41}$$

$$\mathbf{\Gamma} = \begin{bmatrix} 0 & 0 & \gamma_{13} \\ 0 & 0 & \gamma_{23} \\ 0 & 0 & \gamma_{33} \end{bmatrix}, \tag{3.42}$$

$$\mathbf{\Phi} = \begin{bmatrix} \phi_{11} & & \\ \phi_{21} & \phi_{22} & \\ \phi_{31} & \phi_{32} & \phi_{33} \end{bmatrix} = \begin{bmatrix} \sigma_{\xi_1}^2 & & \\ \sigma_{\xi_2\xi_1} & \sigma_{\xi_2}^2 & \\ \sigma_{\xi_3\xi_1} & \sigma_{\xi_3\xi_2} & \sigma_{\xi_3}^2 \end{bmatrix}, \tag{3.43}$$

and

$$\Psi = \begin{bmatrix} \psi_{11} & & \\ \psi_{21} & \psi_{22} & \\ \psi_{31} & \psi_{32} & \psi_{33} \end{bmatrix} = \begin{bmatrix} \sigma_{\zeta_1}^2 & & \\ \sigma_{\zeta_2\zeta_1} & \sigma_{\zeta_2}^2 & \\ \sigma_{\zeta_3\zeta_1} & \sigma_{\zeta_3\zeta_2} & \sigma_{\zeta_3}^2 \end{bmatrix}.$$ (3.44)

2. For the measurement model of the exogenous variables relating the observed variables *Sex*, *MoEd*, *FaEd*, *FaOcc*, and *Sit* to the latent factors *BioSex*, *SES*, and *Situatin*,

$$\Lambda_X = \begin{bmatrix} 1 & 0 & 0 \\ 0 & 1 & 0 \\ 0 & \lambda_{32} & 0 \\ 0 & \lambda_{42} & 0 \\ 0 & 0 & 1 \end{bmatrix}$$ (3.45)

and

$$\Theta_\delta = \begin{bmatrix} 0 & & & & \\ 0 & \theta_{22} & & & \\ 0 & 0 & \theta_{33} & & \\ 0 & 0 & 0 & \theta_{44} & \\ 0 & 0 & 0 & 0 & 0 \end{bmatrix} = \begin{bmatrix} 0 & & & & \\ 0 & \sigma_{\delta_2}^2 & & & \\ 0 & 0 & \sigma_{\delta_3}^2 & & \\ 0 & 0 & 0 & \sigma_{\delta_4}^2 & \\ 0 & 0 & 0 & 0 & 0 \end{bmatrix}.$$ (3.46)

3. For the measurement model of the endogenous variables relating the *HBI* scales T_f, T_c, F_a, F_c, A_t, and A_c to the behavior constructs *Thinking*, *Feeling*, and *Acting*,

$$\Lambda_Y = \begin{bmatrix} \lambda_{11} & \lambda_{12} & 0 \\ 1 & 0 & 0 \\ 0 & \lambda_{32} & \lambda_{33} \\ 0 & 1 & 0 \\ \lambda_{51} & 0 & \lambda_{53} \\ 0 & 0 & 1 \end{bmatrix}$$ (3.47)

and

$$\Theta_\varepsilon = \begin{bmatrix} \theta_{11} & & & & & \\ 0 & \theta_{22} & & & & \\ 0 & 0 & \theta_{33} & & & \\ 0 & 0 & 0 & \theta_{44} & & \\ 0 & 0 & 0 & 0 & \theta_{55} & \\ 0 & 0 & 0 & 0 & 0 & \theta_{66} \end{bmatrix} = \begin{bmatrix} \sigma_{\varepsilon_1}^2 & & & & & \\ 0 & \sigma_{\varepsilon_2}^2 & & & & \\ 0 & 0 & \sigma_{\varepsilon_3}^2 & & & \\ 0 & 0 & 0 & \sigma_{\varepsilon_4}^2 & & \\ 0 & 0 & 0 & 0 & \sigma_{\varepsilon_5}^2 & \\ 0 & 0 & 0 & 0 & 0 & \sigma_{\varepsilon_6}^2 \end{bmatrix}.$$ (3.48)

Counting the free elements in these matrices [equations (3.41)–(3.48)] yields $p = 32$ parameters to be estimated. Since there are $c =$

$(NX + NY)(NX + NY + 1)/2 = (5 + 6)(5 + 6 + 1)/2 = 66$ non-redundant variances and covariances available, the structural equation model in Figure 3.5 probably is overidentified with $df = c - p = 34$ degrees of freedom.

Estimation, Data Fit, and Modification

After a suitable structural equation model has been conceptualized, its basic matrices specified, and its identification status checked, the estimation of unknown parameters can be accomplished by a variety of iterative methods, including those previously discussed, i.e., maximum likelihood (ML) and generalized least squares (GLS). Once estimation has been completed, fit assessments should be conducted to uncover potential data-model inconsistencies. To this end, all discussions in Chapter 2 regarding the fit of a particular data set to a hypothesized CFA model readily extend and apply to the more general models introduced in this chapter. In particular, data-model fit assessments should be based on multiple indices selected from, for example (but not limited to), (a) the plausibility of individual parameter estimates and associated statistics; (b) the chi-square value and chi-square over degrees of freedom ratio; (c) the Goodness-of-Fit Index (GFI) and Adjusted Goodness-of-Fit Index (AGFI); (d) the Normed Fit Index (NFI), Nonnormed Fit Index (NNFI), and Comparative Fit Index (CFI); and/or (e) the Parsimony Goodness-of-Fit Index (PGFI) and Parsimony Normed Fit Index (PNFI). However, recall from Chapter 2 that these indices depend on the multivariate normality assumption and the sample size used; thus, results should be interpreted with caution if underlying assumptions are violated and/or analyses of small data sets are conducted.

If data-model inconsistencies are discovered, the guidelines presented in Chapter 2 should be followed before a model is modified and reestimated to improve the fit results. That is, modifications to a general structural equation model, first and foremost, must make theoretical/substantive sense and *might* be justifiable if (a) a very large chi-square value—as compared to the degrees of freedom—indicates overall data-model misfit; (b) large modification indices (MI)—or, equivalently, Lagrange multipliers (LM)—and associated large expected parameter change statistics (EPC) point to fixed parameters that should be estimated; and/or (c) Wald tests indicate an overfitted model that includes free parameters that could be fixed without significantly eroding the overall data-model fit. Furthermore, whenever possible, results based on a modified structure should be cross-validated with a new and independent sample. If this is not feasible, a cross-validation index (CVI) can be computed based on a random split of the sample or, if the sample size is too small, an estimate of the expected value of the cross-validation index (ECVI) can be used to judge the predictive validity of the model. The following final LISREL

Table 3.6. A Possible LISREL Input File for the Model in Figure 3.5

1	Example 3.2. A Structural Equation Model of Sex, SES, and Situation on T, F, and A
2	DA NI = 11 NO = 167
3	LA
4	Tf Tc Fa Fc At Ac Sex MoEd FaEd FaOcc Sit
5	KM SY
6	1
7	.153 1
8	−.773 −.157 1
9	−.447 .579 .332 1
10	−.106 .054 −.320 .142 1
11	.083 .704 −.310 .487 .354 1
12	−.213 −.003 .086 .188 .136 .056 1
13	.042 .009 −.012 −.059 .036 .031 .052 1
14	−.041 .011 −.026 −.022 .061 .025 .081 .508 1
15	.054 .077 .052 .034 .056 .057 −.011 .363 .526 1
16	−.323 −.176 .495 .096 −.291 −.276 .004 −.046 −.020 −.083 1
17	SD
18	.660 .443 .684 .493 .533 .529 .500 1.991 2.059 1.578 .501
19	MO NY = 6 NE = 3 NX = 5 NK = 3 c
20	LY = FU,FI TE = DI LX = FU,FI TD = DI c
21	BE = ZE GA = FU,FI PH = SY,FR PS = SY,FR
22	LE
23	Thinking Feeling Acting
24	LK
25	BioSex SES Situatin
26	FR GA(1,3) GA(2,3) GA(3,3)
27	VA 1.0 LY(2,1) LY(4,2) LY(6,3)
28	FR LY(1,1) LY(1,2) LY(3,2) LY(3,3) LY(5,1) LY(5,3)
29	VA 1.0 LX(1,1) LX(2,2) LX(5,3)
30	FR LX(3,2) LX(4,2)
31	FI TD(1,1) TD(5,5)
32	OU ME = GLS AD = off SC ND = 3

and EQS examples illustrate some of the above remarks by analyzing the data in Appendix E using the hypothesized model shown in Figure 3.5:

LISREL EXAMPLE 3.2: A STRUCTURAL EQUATION MODEL OF THE EFFECT OF SITUATION SPECIFICITY ON THINKING, FEELING, AND ACTING

LISREL Input. The input file used for the analysis is shown in Table 3.6. All syntax should be familiar to the reader with the exception of two options on the OUtput line (line 32). The statement "ME = GLS" indicates that the chosen Method of Estimation is generalized least squares while the statement

"AD = off" turns off LISREL's ADmissibility check. This internal check terminates the program run after 20 iterations to prevent a nonconverging iteration process from going on indefinitely. According to Jöreskog and Sörbom (1993a), models with solutions that do not converged after 20 iterations are probably incorrect and should be examined for possible mis-specifications. There are, however, situations when this check should be turned off: For example, when diagonal elements in the variance/covariance matrices Θ_ε or Θ_δ are intentionally fixed to 0 (e.g., the variances of the measurement errors of *Sex* and *Sit* in the present model), solutions might not converge unless the number of iterations is increased or the check is turned off all together.

Selected LISREL Output. Table 3.7 presents a partial output file from the above analysis including the unstandardized GLS parameter estimates, reliability estimates of the observed variables, coefficients of determination for the structural equations associated with the latent constructs *Thinking*, *Feeling*, and *Acting*, and goodness-of-fit indices. First, note that none of the estimates appear to have theory-contradicting signs or magnitudes (e.g., none of the variance estimates in **PH**, **PS**, **TD**, and **TE** are negative) and that all overall fit measures indicate no data-model inconsistencies (e.g., $\chi^2 = 41.737$, $df = 34$, $p = 0.170$; NFI, NNFI, and CFI all above 0.99; and PNFI = 0.616). These results answer the first research question posed; that is, they indicate that the *Thinking*, *Feeling*, and *Acting* constructs—as measured by the six *HBI* scales—seem not to depend on the individual's biological sex and socioeconomic status (of course, keep in mind that the present sample of university residence-hall counselors was drawn from a population of rather homogeneous socioeconomic background).

With regard to the second research question, consult the estimates and associated test statistics in the **GA** matrix. They reveal that, based on the present sample, situation specificity (*Situatin*) significantly effects the endogenous constructs *Thinking* ($\gamma_{13} = -0.144$, $t = -2.123$) and *Acting* ($\gamma_{33} = -0.295$, $t = -4.089$) but not *Feeling* ($\gamma_{23} = 0.058$, $t = 0.865$). Specifically, while the question, "How do I view myself as a student?" leads to higher thinking and acting orientations than, "How do I view myself when confronted with a close friend in emotional distress?", somewhat surprisingly, the two questions did not lead to significant differences in the feeling component of the TFA trichotomy.

Finally, note from the estimates and test statistics in the **PH** matrix that all associations between the latent exogenous variables (*BioSex*, *SES*, and *Situatin*) are nonsignificant. Furthermore, measurement error variances of the observed exogenous variable *FaEd* and the observed endogenous *HBI* scales T_f, T_c, and F_a are nonsignificant, which corresponds to the high reliability estimates of 0.912, 0.951, 0.901, and 0.977, respectively [but note the relatively low reliabilities of *MoEd* (0.301) and the *HBI* scale A_t (0.560)].

Table 3.7. Selected Output from the LISREL Analysis of the Model in
Figure 3.5

LISREL ESTIMATES (GENERALIZED LEAST SQUARES)

LAMBDA-Y

	Thinking	Feeling	Acting
Tf	1.884	−2.334	—
	(0.221)	(0.224)	
	8.505	−10.406	
Tc	1.000	—	—
Fa	—	2.211	−1.952
		(0.204)	(0.190)
		10.840	−10.262
Fc	—	1.000	—
At	−1.848	—	2.026
	(0.292)		(0.270)
	−6.337		7.515
Ac	—	—	1.000

LAMBDA-X

	BioSex	SES	Situatin
Sex	1.000	—	—
MoEd	—	1.000	—
FaEd	—	1.807	—
		(0.358)	
		5.048	
FaOcc	—	0.870	—
		(0.141)	
		6.161	
Sit	—	—	1.000

GAMMA

	BioSex	SES	Situatin
Thinking	—	—	−0.144
			(0.068)
			−2.123
Feeling	—	—	0.058
			(0.067)
			0.865
Acting	—	—	−0.295
			(0.072)
			−4.089

Table 3.7 (*cont.*)

PHI		
	BioSex	SES	Situatin
BioSex	0.213 (0.025) 8.413		
SES	0.037 (0.043) 0.869	1.146 (0.353) 3.246	
Situatin	−0.007 (0.019) −0.345	−0.014 (0.044) −0.314	0.245 (0.027) 9.008

PSI		
	Thinking	Feeling	Acting
Thinking	0.169 (0.023) 7.388		
Feeling	0.124 (0.019) 6.657	0.151 (0.025) 6.123	
Acting	0.155 (0.021) 7.553	0.125 (0.021) 6.074	0.173 (0.027) 6.311

SQUARED MULTIPLE CORRELATIONS FOR
STRUCTURAL EQUATIONS

Thinking	Feeling	Acting
0.029	0.005	0.109

THETA-EPS

Tf	Tc	Fa	Fc	At	Ac
0.019 (0.018) 1.060	0.019 (0.010) 1.866	0.010 (0.017) 0.593	0.068 (0.010) 6.738	0.119 (0.020) 5.849	0.081 (0.013) 6.502

SQUARED MULTIPLE CORRELATIONS FOR Y-VARIABLES

Tf	Tc	Fa	Fc	At	Ac
0.951	0.901	0.977	0.690	0.560	0.705

THETA-DELTA

Sex	MoEd	FaEd	FaOcc	Sit
—	2.664 (0.344) 7.747	0.362 (0.611) 0.593	1.259 (0.208) 6.051	—

Table 3.7 (*cont.*)

SQUARED MULTIPLE CORRELATIONS FOR X-VARIABLES				
Sex	MoEd	FaEd	FaOcc	Sit
1.000	0.301	0.912	0.408	1.000

GOODNESS OF FIT STATISTICS

CHI-SQUARE WITH 34 DEGREES OF FREEDOM = 41.737 (P = 0.170)

GOODNESS OF FIT INDEX (GFI) = 0.954
ADJUSTED GOODNESS OF FIT INDEX (AGFI) = 0.911
PARSIMONY GOODNESS OF FIT INDEX (PGFI) = 0.492

NORMED FIT INDEX (NFI) = 0.996
NON-NORMED FIT INDEX (NNFI) = 0.999
PARSIMONY NORMED FIT INDEX (PNFI) = 0.616
COMPARATIVE FIT INDEX (CFI) = 0.999

EQS EXAMPLE 3.2: A STRUCTURAL EQUATION MODEL OF THE EFFECT OF SITUATION SPECIFICITY ON THINKING, FEELING, AND ACTING. For reference purposes, the structural equation model of Figure 3.5 is reproduced in Figure 3.6 in the Bentler-Weeks notation. The following analysis is based on the same data set as the previous LISREL example; the only aspects of the analysis that are different from those in the previous one are discussed in the following.

EQS Input. The EQS input file for an analysis of the model in Figure 3.6 is shown in Table 3.8. At this point, the only unfamiliar statement should be the "Set = GFF;" subcommand of the /LMTEST section on line 53. Recall from Chapter 2, EQS Example 2.1, that the user can request Lagrange Multiplier tests for specific parameters with the "Set" command. Here, LM tests are computed for the fixed structural relations among the latent factors, F (i.e., the hypothesized 0 effects from *BioSex* and *SES* to *Thinking*, *Feeling*, and *Acting* that, in the Bentler-Weeks model, are specified in the **GAMMA** matrix). More specifically, the statement "Set = GFF;" requests LM tests for fixed coefficients among Factors (indicated by the letter combination FF) in the **GAMMA** matrix (indicated by the letter G preceding the FF combination).

Selected EQS Output. A partial EQS output file from the analysis of the model in Figure 3.6 is reproduced in Table 3.9. Only the standardized solutions and the results of the Wald and Lagrange Multiplier tests are shown since the unstandardized parameter estimates and the data-model fit results from the EQS analysis are the same as those obtained in the previous LISREL example (see Table 3.7). Also, the focus is not on the *HBI* measurement model since it was discussed in Chapter 2. However, recall from LISREL and EQS Example 2.2—and note in Table 3.9—that each of the

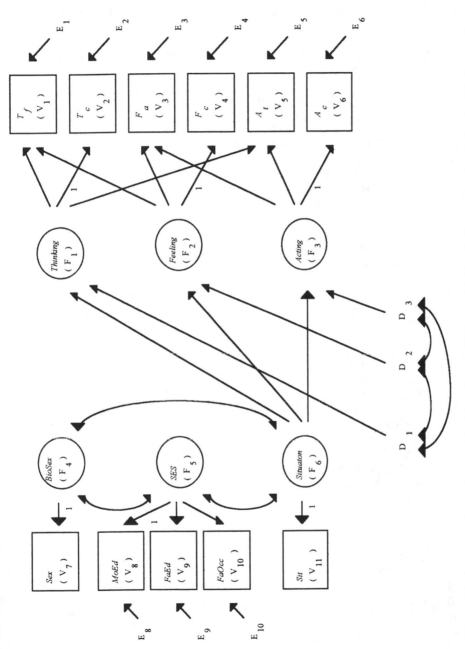

Figure 3.6. A structural equation model of situation specificity on thinking, feeling, and acting in Bentler-Weeks notation.

Table 3.8. An EQS Input File for the Model in Figure 3.6

1	/TITLE
2	Example 3.2. A Structural Equation Model of Sex, SES, and Situation on T, F, and A
3	/SPECIFICATIONS
4	Variables = 11; Cases = 167; Matrix = Correlation; Analysis = Covariance; Method = GLS;
5	/LABELS
6	V1 = Tf; V2 = Tc; V3 = Fa; V4 = Fc; V5 = At; V6 = Ac;
7	V7 = Sex; V8 = MoEd; V9 = FaEd; V10 = FaOcc; V11 = Sit;
8	F1 = Thinking; F2 = Feeling; F3 = Acting;
9	F4 = BioSex; F5 = SES; F6 = Situatin;
10	/EQUATIONS
11	V1 = *F1 + *F2 + E1;
12	V2 = F1 + E2;
13	V3 = *F2 + *F3 + E3;
14	V4 = F2 + E4;
15	V5 = *F1 + *F3 + E5;
16	V6 = F3 + E6;
17	V7 = F4 + E7;
18	V8 = F5 + E8;
19	V9 = *F5 + E9;
20	V10 = *F5 + E10;
21	V11 = F6 + E11;
22	F1 = *F6 + D1;
23	F2 = *F6 + D2;
24	F3 = *F6 + D3;
25	/VARIANCES
26	F4 to F6 = *;
27	E1 to E6 = *;
28	E8 to E10 = *;
29	E7, E11 = 0;
30	D1 to D3 = *;
31	/COVARIANCES
32	F6, F4 = *;
33	F6, F5 = *;
34	F5, F4 = *;
35	D3, D1 = *;
36	D3, D2 = *;
37	D2, D1 = *;
38	/MATRIX
39	1
40	.153 1
41	−.773 −.157 1
42	−.447 .579 .332 1
43	−.106 .054 −.320 .142 1
44	.083 .704 −.310 .487 .354 1
45	−.213 −.003 .086 .188 .136 .056 1

Table 3.8 (*cont.*)

46	.042 .009 −.012 −.059 .036 .031 .052 1
47	−.041 .011 −.026 −.022 .061 .025 .081 .508 1
48	.054 .077 .052 .034 .056 .057 −.011 .363 .526 1
49	−.323 −.176 .495 .096 −.291 −.276 .004 −.046 −.020 −.083 1
50	/STANDARD DEVIATIONS
51	.660 .443 .684 .493 .533 .529 .500 1.991 2.059 1.578 .501
52	/LMTEST
53	Set = GFF;
54	/WTEST
55	/PRINT
56	Digit = 4;
57	/END

equations for the *HBI* comparison scales T_f, F_a, and A_t contain two standardized coefficients larger than 1.00, so that some caution should be exercised in interpreting the results.

The standardized solutions indicate that *FaEd* seems to be the most valid indicator of the latent construct *SES* [as is indicated by the largest standardized factor loading (0.9548) compared to those associated with *MoEd* (0.5485) and *FaOcc* (0.6389)]. Further, situation specificity seems to have the strongest direct effect on the TFA construct *Acting* [as is indicated by the largest standardized coefficient (−0.3307) compared to the situation effects on *Thinking* (−0.1711) and Feeling (0.0742)].

The results from the Wald tests indicate those free parameters in the model that could be fixed to 0 without significantly eroding the fit of the model. Here, seven parameters—the structural effect from *Situatin* to *Feeling*; the three covariances between the exogenous constructs *BioSex*, *SES*, and *Situatin*; and the variances of the measurement errors associated with the indicator variables *FaEd*, F_a, and T_f—could be fixed without significantly increasing the overall chi-square statistic. Note that these parameters are the ones that were identified in the previous LISREL example as being nonsignificant.

On the other hand, the Lagrange Multiplier tests indicate which fixed parameters in the model could be freed to increase the overall data-model fit. Specifically, the multivariate LM statistics show that, if the direct structural effect from *BioSex* to the TFA construct *Feeling* would be freed during a subsequent analysis, the (already nonsignificant) overall chi-square statistic of 41.737 ($df = 34$, $p = 0.170$; see Table 3.7) would decrease significantly.

In conclusion, the Wald and Lagrange Multiplier tests point to several possible modifications of the hypothesized model in Figure 3.6. However, since the present data seem to fit the proposed structure and the two research questions could be answered by an analysis of the initial model, no modifications to the model seem necessary or justified (but see Exercise 3.5).

Table 3.9. Selected EQS Output from an Analysis of the Model in Figure 3.6

STANDARDIZED SOLUTION:
GENERALIZED LEAST SQUARES SOLUTION (NORMAL DISTRIBUTION THEORY)

TF = V1 =   1.2621*F1 −   1.4585*F2 +   .2205 E1
TC = V2 =   .9491 F1 +   .3149 E2
FA = V3 =   1.2833*F2 −   1.2822*F3 +   .1516 E3
FC = V4 =   .8308 F2 +   .5565 E4
AT = V5 =   −1.4830*F1 +   1.7176*F3 +   .6630 E5
AC = V6 =   .8395 F3 +   .5433 E6
SEX = V7 =   1.0000 F4 +   .0000 E7
MOED = V8 =   .5485 F5 +   .8362 E8
FAED = V9 =   .9548*F5 +   .2973 E9
FAOCC = V10 =   .6389*F5 +   .7693 E10
SIT = 11 =   1.0000 F6 +   .0000 E11
THINKING = F1 = −.1711*F6 +   .9852 D1
FEELING = F2 =   .0742*F6 +   .9972 D2
ACTING = F3 = −.3307*F6 +   .9437 D3

CORRELATIONS AMONG INDEPENDENT VARIABLES

      V   F
      _   _

                    I   F5 -SES   .0752*   I
                    I   F4 -BIOSEX   I
                    I   I
                    I   F6 -SITUATIN   −.0288*I
                    I   F4 -BIOSEX   I
                    I   I
                    I   F6 -SITUATIN   −.0259*I
                    I   F5 -SES   I
                    I   I

CORRELATIONS AMONG INDEPENDENT VARIABLES

E		D		
	I	D2-FEELING	.7783*	I
	I	D1-THINKING		I
	I			I
	I	D3-ACTING	9079*	I
	I	D1-THINKING		I
	I			I
	I	D3-ACTING	.7732*	I
	I	D2-FEELING		I
	I			I

WALD TEST (FOR DROPPING PARAMETERS)
MULTIVARIATE WALD TEST BY SIMULTANEOUS PROCESS
CUMULATIVE MULTIVARIATE STATISTICS

STEP	PARAMETER	CHI-SQUARE	D.F.	PROBABILITY	UNIVARIATE INCREMENT	
					CHI-SQUARE	PROBABILITY
1	F6,F5	0.098	1	0.754	0.098	0.754
2	F6,F4	0.200	2	0.905	0.101	0.750
3	E9,E9	0.535	3	0.911	0.336	0.562
4	E3,E3	0.912	4	0.923	0.377	0.539
5	F5,F4	1.534	5	0.909	0.622	0.430
6	F2,F6	2.407	6	0.879	0.873	0.350
7	E1,E1	5.443	7	0.606	3.036	0.081

Table 3.9 (cont.)

LAGRANGE MULTIPLIER TEST (FOR ADDING PARAMETERS)

ORDERED UNIVARIATE TEST STATISTICS:

NO	CODE	PARAMETER	CHI-SQUARE	PROBABILITY	PARAMETER CHANGE
1	2 16	F2,F4	5.471	0.019	0.107
2	2 16	F1,F4	4.391	0.036	−0.073
3	2 16	F3,F4	0.821	0.365	0.031
4	2 16	F3,F5	0.531	0.466	0.011
5	2 16	F1,F5	0.473	0.492	−0.010
6	2 16	F2,F5	0.022	0.882	0.003
7	2 0	V11,F6	0.000	1.000	0.000
8	2 0	V7,F4	0.000	1.000	0.000
9	2 0	F2,D2	0.000	1.000	0.000
10	2 0	V8,F5	0.000	1.000	0.000
11	2 0	F1,D1	0.000	1.000	0.000
12	2 0	F3,D3	0.000	1.000	0.000
13	2 0	V2,F1	0.000	1.000	0.000
14	2 0	V4,F2	0.000	1.000	0.000
15	2 0	V6,F3	0.000	1.000	0.000

MULTIVARIATE LAGRANGE MULTIPLIER TEST BY SIMULTANEOUS PROCESS IN STAGE 1

PARAMETER SETS (SUBMATRICES) ACTIVE AT THIS STAGE ARE:

GFF

CUMULATIVE MULTIVARIATE STATISTICS

					UNIVARIATE INCREMENT	
STEP	PARAMETER	CHI-SQUARE	D.F.	PROBABILITY	CHI-SQUARE	PROBABILITY
1	F2,F4	5.471	1	0.019	5.471	0.019

Conclusion

Structural equation modeling, as introduced throughout this book and reviewed and summarized in the last section, should be understood as a research process, from model conceptualization and specification, identification, estimation, and data-model fit assessment, to possible model modification. Results from an SEM analysis can be of great value only if the hypothesized structures are built on a strong substantive/theoretical foundation. Obviously, it must remain the user's responsibility to conceptualize meaningful, theory-based models if the applications of the statistical tools introduced here are to lead to a deeper understanding of the modeled social science phenomena.

A conceptual structural equation model (a path analytical structure with or without latent variables or a confirmatory factor analysis model) can be represented mathematically by the following three basic matrix equations:

1. An equation relating the endogenous latent variables η_r to the exogenous latent variables ξ_s,

$$\eta = B\eta + \Gamma\xi + \zeta,$$

where η is a $(NE \times 1)$ column vector of endogenous latent variables, ξ is a $(NK \times 1)$ column vector of exogenous latent variables, B is a $(NE \times NE)$ matrix of structural coefficients from endogenous to other endogenous latent variables, Γ is a $(NE \times NK)$ matrix of structural coefficients from exogenous to endogenous latent variables, and ζ is a $(NE \times 1)$ column vector of error terms associated with the endogenous latent variables.

2. An equation relating the exogenous observed variables X_j to the exogenous latent constructs ξ_s,

$$X = \Lambda_x\xi + \delta,$$

where X is a $(NX \times 1)$ column vector of observed variables measured as deviations from their means, Λ_x is a $(NX \times NK)$ matrix of structural coefficients from latent exogenous to their observed indicator variables, ξ is a $(NK \times 1)$ column vector of latent exogenous constructs, and δ is a $(NX \times 1)$ column vector of measurement error terms of the observed variables.

3. An equation relating the endogenous observed variables Y_i to the exogenous latent constructs η_r,

$$Y = \Lambda_Y\eta + \varepsilon,$$

where Y is a $(NY \times 1)$ column vector of observed variables measured as deviations from their means, Λ_Y is a $(NY \times NE)$ matrix of structural coefficients from latent endogenous variables to their observed indicators, η is a $(NE \times 1)$ column vector of latent endogenous constructs, and ε is a $(NY \times 1)$ column vector of measurement error terms of the observed variables.

In addition to the four structural matrices $\mathbf{B}$, $\mathbf{\Gamma}$, $\mathbf{\Lambda}_X$, and $\mathbf{\Lambda}_Y$, a set of variance/covariance matrices needs to be specified to completely determine the hypothesized model: (a) a $(NK \times NK)$ variance/covariance matrix $\mathbf{\Phi}$ of the exogenous latent variables, (b) a $(NE \times NE)$ variance/covariance matrix $\mathbf{\Psi}$ of error terms associated with the model-implied structural equations, (c) a $(NX \times NX)$ variance/covariance matrix $\mathbf{\Theta}_\delta$ of measurement errors of the observed exogenous variables, and (d) a $(NY \times NY)$ variance/covariance matrix $\mathbf{\Theta}_\varepsilon$ of measurement error terms associated with the observed endogenous variables.

The identification of a structural equation model is an a priori condition for the estimation of unknown parameters based on a particular data set. Once identification is ensured, the user can choose from among several techniques available in the LISREL and EQS packages that produce consistent parameter estimates. When the maximum likelihood (ML) or generalized least squares (GLS) methods are used, a multivariate normal distribution of the data must be assumed; otherwise, the arbitrary distribution free (ADF) technique should be utilized. All three methods are iterative in nature; that is, a particular fitting function of the unrestricted and model-implied variance/covariance matrix of the observed variables is minimized to produce the parameter estimates.

For overidentified structures, the minimum value of the fitting function can be used as a basis for the evaluation of data-model consistency (just-identified models produce a 0-valued fitting function since, then, the unrestricted and model-implied variance/covariance matrices are equal). A multitude of indices have been developed to assist the researcher in determining whether or not the analyzed data set fits the hypothesized model, each with important advantages and disadvantages. Often, however, just a careful examination of the obtained parameter estimates (for example, do estimates contradict the hypothesized theory or any well-known statistical principles?) will go a long way in identifying certain model misspecifications.

Unfortunately, the goal in some applications of structural equation modeling seems to be the identification of the model that best fits the data. To this end, initially hypothesized models often are modified to improve the "model fit." Here, I used the term "data fit" to emphasize that users (especially those new to SEM) should fit the *data* to an a priori hypothesized model and resist the temptation to change a hastily conceived model until it fits the data set under consideration. One of the strengths of SEM lies in its disconfirmatory —not its confirmatory—power: *Based on empirical data, theoretically proposed structures can be rejected as good approximations to "reality" but cannot be confirmed as being the representation of the true underlying processes.* Post-hoc model modifications should be considered only if they make substantive/theoretical sense, and, whenever possible, the modified models should be cross-validated with independent data.

EXERCISES

3.1. Consider the structural equation model shown in Figure 3.1. Run the LISREL program in Table 3.2 after fixing the direct effects of *AcRank* on *ColgPres* and *AcRank* on *SES* to 0.
   (a) Assess whether or not the data fit the proposed model by considering the various fit indices discussed in Chapter 2.
   (b) Based on the results obtained in part (a) and an examination of the modification indices and associated expected parameter change statistics, if appropriate, modify and reanalyze the structure to improve data-model fit.

3.2. Redo Exercise 3.1, this time using data from the female subsample (see Appendix D).
   (a) Compare fit results and modification decisions to those obtained from the analysis of the male sample in Exercise 3.1.
   (b) Based on analyses of the initial model in Figure 3.1, compare the direct, indirect, and total effects of parents' on respondent's socioeconomic status across the male and female subsamples.

3.3. Use the rules and results in Appendix C and the definitions and associated assumptions of the eight basic matrices (**B**, **Γ**, **Φ**, **Ψ**, Λ_X, Θ_δ, Λ_Y, and Θ_ε) to develop the expression of the model-implied variance/covariance matrix $\Sigma(\theta)$ in equation (3.37) from the unrestricted variance/covariance matrix Σ given in equation (3.36).

3.4. Use the input file of Table 3.4 to rerun the EQS analysis of the model in Figure 3.4 and compare the obtained GLS estimates of the indirect and total effects among the five latent variables to the corresponding ML estimates from the LISREL analysis in Table 3.3.

3.5. Modify the structural equation model in Figures 3.3 and 3.4 according to the results from the Wald and Lagrange Multiplier tests shown in Table 3.9 and use either LISREL, SIMPLIS, or EQS to estimate the modified model. Compare the estimated expected values of the cross-validation index (ECVI) from the intitial and the modified models.

3.6. Select a path analysis example from Chapter 1 and a CFA example from Chapter 2 and use either LISREL, SIMPLIS, or EQS to reestimate the parameters with the method of generalized least squares. Compare the obtained results to the original ones presented in the appropriate tables in the text.

3.7. Reconsider Exercise 1.13. Use Appendix D and your theoretical justification of the proposed structure to select multiple indicators (if available) for each of the constructs in the model you conceptualized. Use either LISREL, SIMPLIS, or EQS to complete the following tasks:
   (a) Conduct a confirmatory factor analysis of the exogenous variables. If appropriate, modify the hypothesized model.
   (b) Conduct a CFA of the endogenous variables. Again, if appropriate, modify the proposed structure.
   (c) Combine the final models of parts (a) and (b) into an overidentified general structural equation model mirroring the structure of the model conceptualized in Exercise 1.13. Analyze the proposed model and interpret the results.

Recommended Readings

Various social science research methods books contain chapters on general structural equation modeling. For example, see

Nesselroade, J.R., and Cattell, R.B. (Eds.). (1988). *Handbook of Multivariate Experimental Psychology* (2nd ed.). New York: Plenum Press.
Pedhazur, E.J., and Schmelkin, L. (1991). *Measurement, Design, and Analysis: An Integrated Approach*. Hillsdale, NJ: Lawrence Erlbaum.
Thompson, B. (Ed.). (1989). *Advances in Social Science Methodology* (Vol. 1). Greenwich, CT: JAI Press.

For a historical perspective on early developments and treatments of general SEM methods, consult

Blalock, H.M. (1964). *Causal Inferences in Nonexperimental Research*. Chapel Hill: The University of North Carolina Press.
Blalock, H.M. (Ed.). (1985a). *Causal Models in the Social Sciences* (2nd ed.). New York: Aldine.
Blalock, H.M. (Ed.). (1985b). *Causal Models in Panel and Experimental Designs*. New York: Aldine.
Long, J.S. (1983b). *Covariance Structure Models: An Introduction to LISREL*. Beverly Hills, CA: Sage.

Excellent reference texts that were mentioned throughout the book and are highly recommended, especially to the reader who wants to embark on a more in-depth and comprehensive study of structural equation modeling, are

Bollen, K.A. (1989). *Structural Equations with Latent Variables*. New York: John Wiley & Sons.
Hayduk, L.A. (1987). *Structural Equation Modeling with LISREL: Essentials and Advances*. Baltimore: Johns Hopkins University Press.
Loehlin, J.C. (1992). *Latent Variable Models: An Introduction to Factor, Path, and Structural Analysis* (2nd ed.). Hillsdale, NJ: Lawrence Erlbaum.

The various manuals accompanying the LISREL and EQS packages provide very helpful explanations, references, and examples. These are

Bentler, P.M. (1993). *EQS: Structural Equations Program Manual*. Los Angeles: BMDP Statistical Software.
Bentler, P.M., and Wu, E.J.C. (1993). *EQS/Windows: User's Guide*. Los Angeles: BMDP Statistical Software.
Jöreskog, K.G., and Sörbom, D. (1993a). *LISREL 8 User's Reference Guide*. Chicago: Scientific Software International.
Jöreskog, K.G., and Sörbom, D. (1993b). *LISREL 8: Structural Equation Modeling with the SIMPLIS Command Language*. Chicago: Scientific Software.
Jöreskog, K.G., and Sörbom, D. (1993c). *PRELIS 2: A Program for Multivariate Data Screening and Data Summarization: A Preprocessor for LISREL*. Chicago: Scientific Software International.

Finally, *Structural Equation Modeling: A Multidisciplinary Journal*, published by Lawrence Erlbaum Associates, is a quarterly journal that publishes theoretical as well as applied articles and book and software reviews dealing with all facets of SEM. Most papers are readily accessible to the interested reader familiar with *Basic Concepts of Structural Equation Modeling*.

APPENDIX A

The SIMPLIS Command Language

Overview and Key Points

The SIMPLIS (SIMPle LISrel) command language within the LISREL package gives the user the option of conducting path, confirmatory factor, or full structural equation model analyses without having to specify explicitly the 0 and non-zero elements in each of the basic matrices $\mathbf{B}$, $\mathbf{\Gamma}$, $\mathbf{\Phi}$, $\mathbf{\Psi}$, $\mathbf{\Lambda}_X$, $\mathbf{\Theta}_\delta$, $\mathbf{\Lambda}_Y$, and $\mathbf{\Theta}_\varepsilon$. An English-like syntax is used to easily specify a wide variety of models, and, with the MS Windows version of LISREL, output options include drawings of path diagrams with attached parameter estimates, t-values (the nonsignificant ones are distinguished from the significant ones by being displayed in a different color), modification indices, and expected parameter change statistics. One of the most advanced SIMPLIS options after requesting a path diagram and estimating a model is the possibility of model modification by freeing (or fixing) parameters on-screen through "pointing," "clicking," and "dragging" in the diagram. A pull-down menu then gives the option of reestimating and displaying the modified model. Although very convenient and user-friendly, the researcher should be aware that these options can be abused easily: With an ill-conceived and ill-fitting initial model, it becomes all too tempting to "go fishing" in search of a model—any model —that, by chance, will fit a particular data set. As I have stressed throughout the book, the user of SEM techniques again is urged to conceptualize theoretically sound models *prior to data analysis* and adjust initial models only if the modification is substantively justified. If this is not possible, tools such as exploratory factor analysis could be used to uncover possible structures underlying the variables in the current data set, and, with a different data set, these structures subsequently could be evaluated with the confirmatory methods discussed here.

The tables in this appendix contain the SIMPLIS input files and selected output corresponding to each of the LISREL examples discussed in the book. The reader should consult Jöreskog and Sörbom (1993b) for a detailed description of the SIMPLIS command language. However, before the input files are presented, some key points regarding the SIMPLIS syntax are listed.

1. A typical SIMPLIS program is divided into sections by certain header lines such as OBSERVED VARIABLES, COVARIANCE MATRIX, SAMPLE SIZE, and RELATIONSHIPS. Optionally, each such header can end with a colon (:) to increase readability.

2. The first line in a SIMPLIS program usually is a title line that can contain any information except start with the strings of characters *Observed Variables*, *Labels*, or *DA*. To avoid possible problems, one should start the title line with an exclamation point (!), the character used in LISREL to indicate a comment line (i.e., everything in a line typed after "!" anywhere in the input is ignored by the program).

3. After the title, unique names (up to eight characters in length) *must* be given to the observed variables in a model. These labels can be listed in free format after the SIMPLIS headers OBSERVED VARIABLES or LABELS.

4. Information regarding the input data *must* be given next. SIMPLIS accepts raw data or a covariance or correlation matrix together with means and/or standard deviations. Correspondingly, appropriate header lines are RAW DATA, COVARIANCE MATRIX, CORRELATION MATRIX, MEANS, and/or STANDARD DEVIATIONS.

5. After specification of the input data, the sample size (n) is given following the header SAMPLE SIZE.

6. Observed variables may be reordered to increase the readability of the output by listing the variables in their new order after the key words *Reorder Variables*.

7. If the model contains latent variables, they are identified by descriptive labels (up to eight characters in length; different from those for the observed variables) after the header LATENT VARIABLES or UNOBSERVED VARIABLES.

8. The section entitled RELATIONSHIPS (or RELATIONS or EQUATIONS) contains all model-implied equations linking observed and latent variables. The general format of a statement in this section is

dependent (latent or observed) variable(s)

= independent (latent or observed) variable(s)

Structural coefficients linking a dependent to an independent variable can be fixed to a constant by writing the constant—followed by an asterisk (*)—in front of the appropriate independent variable. For example, if *MoEd* is one of three indicators of the latent variable *PaSES*, the unit of measurement of the independent variable *PaSES* can be set equal to that of *MoEd* by the statement

MoEd = 1*PaSES

9. If no reference variables are specified for the purpose of assigning a unit of measurement to the latent variables, SIMPLIS assumes that the latent variables are standardized to unit variance.

10. All measurement error terms of observed variables are free parameters by default. The user can override this default and specify an error variance for a variable, *Var*, to equal some value, *a*, with the statement

    Let the Error Variance of Var *be* a

or

    Set the Error Variance of Var *equal to* a.

11. Covariances between any error terms in a model are 0 by default. However, covariances between (a) measurement errors δ of observed exogenous variables X, (b) measurement errors ε of observed endogenous variables Y, and (c) disturbance terms ζ of latent endogenous variables η can be set free by statements of the form

    Let the Errors between VarA *and* VarB *Correlate*

or

    Set the Error Covariance between VarA *and* VarB *Free.*

12. The latent exogenous variables ξ are assumed to be correlated. To override this default, specify, for example,

    Set the Covariances of Ksi1 − Ksi2 *to* 0

or

    Set the Correlation of Ksi1 − Ksi2 *to* 0.

13. Various options such as the estimation method, number of decimals printed in the output, or the maximum number of iterations can be specified with the key words *Method, Number of Decimals*, and *Iterations*, respectively.

14. A graphic representation of an estimated model [and access to the advanced features mentioned above (e.g., on-screen model modification)] can be obtained by specifying PATH DIAGRAM in a SIMPLIS input file.

15. When using the SIMPLIS command language, one still can obtain the traditional LISREL output by including the header LISREL OUTPUT in the SIMPLIS program. Now all LISREL output options such as SC (Standardized Completely) or EF (total and indirect EFfects) are available.

16. The optional header END OF PROBLEM indicates the end of the input file.

Table A.1. SIMPLIS Input File for the Simple Linear
Regression in Example 1.1

1	!Example 1.1. SIMPLIS: Simple Linear Regression
2	OBSERVED VARIABLES: Degree FaEd
3	CORRELATION MATRIX:
4	1
5	.129 1
6	MEANS:
7	4.535 3.747
8	STANDARD DEVIATIONS:
9	.962 1.511
10	SAMPLE SIZE: 3094
11	RELATIONSHIPS:
12	Degree = FaEd
13	Number of Decimals = 3
14	END OF PROBLEM

Table A.1(a). Partial SIMPLIS Output from the Simple Linear Regression
in Example 1.1

LISREL ESTIMATES (MAXIMUM LIKELIHOOD)

Degree =	4.227	+	0.0821*FaEd,	Errorvar. =	0.910,	$R^2 = 0.0166$
	(0.0459)		(0.0114)		(0.0231)	
	92.153		7.234		39.319	

Table A.2. SIMPLIS Input File for the Multiple Linear
Regression in Example 1.2

1	!Example 1.2. SIMPLIS: Multiple Linear Regression
2	OBSERVED VARIABLES: Degree FaEd DegreAsp Selctvty
3	COVARIANCE MATRIX:
4	.925
5	.188 2.283
6	.247 .187 1.028
7	.486 .902 .432 3.960
8	MEANS:
9	4.535 3.747 4.003 5.016
10	SAMPLE SIZE: 3094
11	RELATIONSHIPS:
12	Degree = FaEd DegreAsp Selctvty
13	Number of Decimals = 3
14	END OF PROBLEM

Table A.2(a). Partial SIMPLIS Output from the Multiple Linear Regression in Example 1.2

LISREL ESTIMATES (MAXIMUM LIKELIHOOD)

Degree =   3.170   + 0.0289*FaEd + 0.195*DegreAsp + 0.0949*Selctvty,
             (0.0768)   (0.0114)   (0.0165)   (0.00876)
             41.288   2.543   11.804   10.823

             Errorvar. =   0.825, $R^2 = 0.108$
                            (0.0210)
                            39.306

Table A.3. SIMPLIS Input File for the Path Analysis Model in Figure 1.1, Example 1.3

1	!Example 1.3. SIMPLIS: Path Analysis With One Exogenous Variable
2	OBSERVED VARIABLES: Degree FaEd DegreAsp Selctvty
3	COVARIANCE MATRIX:
4	.925
5	.188 2.283
6	.247 .187 1.028
7	.486 .902 .432 3.960
8	SAMPLE SIZE: 3094
9	Reorder Variables: DegreAsp Selctvty Degree FaEd
10	RELATIONSHIPS:
11	DegreAsp = FaEd
12	Selctvty = FaEd DegreAsp
13	Degree = FaEd DegreAsp Selctvty
14	LISREL OUTPUT: SC ND = 3
15	END OF PROBLEM

Table A.3(a). Partial SIMPLIS Output from the Analysis
of the Model in Figure 1.1

LISREL ESTIMATES (MAXIMUM LIKELIHOOD)
  BETA

	DegreAsp	Selctvty	Degree
DegreAsp	—	—	—
Selctvty	0.354	—	—
	(0.033)		
	10.612		
Degree	0.195	0.095	—
	(0.017)	(0.009)	
	11.808	10.827	

  GAMMA

	FaEd
DegreAsp	0.082
	(0.012)
	6.839
Selctvty	0.366
	(0.022)
	16.374
Degree	0.029
	(0.011)
	2.543

  PHI

FaEd
2.283

  PSI

DegreAsp	Selctvty	Degree
1.013	3.477	0.825
(0.026)	(0.088)	(0.021)
39.319	39.319	39.319

SQUARED MULTIPLE CORRELATIONS FOR
  STRUCTURAL EQUATIONS

DegreAsp	Selctvty	Degree
0.015	0.122	0.108

Table A.4. SIMPLIS Input File for the Path Analysis Model in Figure 1.6, Example 1.4

1	!Example 1.4. SIMPLIS: Path Analysis With Two Exogenous Variables
2	OBSERVED VARIABLES: DegreAsp Selctvty Degree FaEd HSRank
3	CORRELATION MATRIX:
4	1
5	.214 1
6	.253 .254 1
7	.122 .300 .129 1
8	.194 .372 .189 .128 1
9	STANDARD DEVIATIONS:
10	1.014 1.990 .962 1.511 .777
11	SAMPLE SIZE: 3094
12	RELATIONSHIPS:
13	DegreAsp = FaEd HSRank
14	Selctvty = FaEd HSRank DegreAsp
15	Degree = FaEd HSRank DegreAsp Selctvty
16	PATH DIAGRAM
17	LISREL OUTPUT: SC EF ND = 3
18	END OF PROBLEM

Table A.4(a). SIMPLIS PATH DIAGRAM Output from an Analysis of the
Model in Figure 1.6

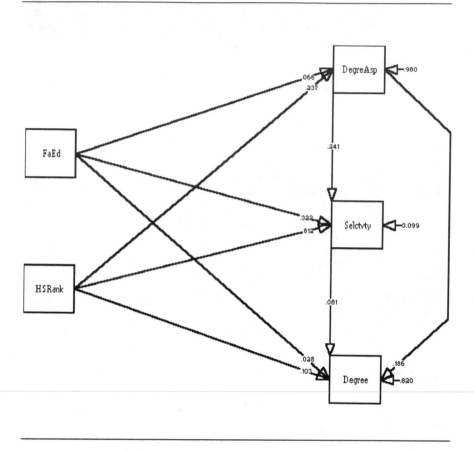

Table A.5. SIMPLIS Input File for the Overidentified
Model in Figure 1.10, Example 1.5

1	!Example 1.5. SIMPLIS: An Over-Identified Model
2	OBSERVED VARIABLES: DegreAsp Degree FaEd
3	COVARIANCE MATRIX:
4	1.028
5	.247 .925
6	.187 .188 2.283
7	SAMPLE SIZE: 3094
8	RELATIONSHIPS:
9	DegreAsp = FaEd
10	Degree = DegreAsp
11	Number of Decimals = 3
12	END OF PROBLEM

Table A.5(a). Partial SIMPLIS Output from an Analysis of the Model in
Figure 1.10

LISREL ESTIMATES (MAXIMUM LIKELIHOOD)

DegreAsp =	0.0819*FaEd, (0.0120) 6.839	Errorvar. =	1.013, (0.0258) 39.319	$R^2 = 0.0149$
Degree =	0.240*DegreAsp, (0.0165) 14.560	Errorvar. =	0.866, (0.0220) 39.319	$R^2 = 0.0642$

GOODNESS OF FIT STATISTICS

CHI-SQUARE WITH 1 DEGREE OF FREEDOM = 32.691 (P = 0.0)

Table A.6. SIMPLIS Input File for the CFA Model in Figure 2.1, Example 2.1

1	!Example 2.1. SIMPLIS: CFA of Parents' SES and Academic Rank
2	OBSERVED VARIABLES: MoEd FaEd PaJntInc HSRank
3	CORRELATION MATRIX:
4	1
5	.610 1
6	.446 .531 1
7	.115 .128 .055 1
8	STANDARD DEVIATIONS:
9	1.229 1.511 2.649 .777
10	SAMPLE SIZE: 3094
11	LATENT VARIABLES: PaSES AcRank
12	RELATIONSHIPS:
13	MoEd = 1*PaSES
14	FaEd PaJntInc = PaSES
15	HSRank = 1*AcRank
16	Set the Error Variance of HSRank to 0
17	Number of Decimals = 3
18	END OF PROBLEM

Table A.6(a). Partial SIMPLIS Output from an Analysis of the Model in Figure 2.1

LISREL ESTIMATES (MAXIMUM LIKELIHOOD)

MoEd =	1.000*PaSES,	Errorvar. =	0.737, (0.0285) 25.827	$R^2 = 0.512$
FaEd =	1.467*PaSES, (0.0483) 30.355	Errorvar. =	0.618, (0.0488) 12.681	$R^2 = 0.729$
PaJntInc =	1.870*PaSES, (0.0628) 29.796	Errorvar. =	4.312, (0.133) 32.361	$R^2 = 0.386$
HSRank =	1.000*AcRank,			$R^2 = 1.000$

COVARIANCE MATRIX OF INDEPENDENT VARIABLES

	PaSES	AcRank
PaSES	0.774 (0.040) 19.419	
AcRank	0.098 (0.014) 7.055	0.604 (0.015) 39.326

Table A.7. SIMPLIS Input File for the *HBI* Model in Figure 2.6, Example 2.2

1	!Example 2.2. SIMPLIS: Validity and Reliability of the HBI
2	OBSERVED VARIABLES: Tf Tc Fa Fc At Ac
3	COVARIANCE MATRIX:
4	.436
5	.045 .196
6	−.349 −.048 .468
7	−.145 .126 .112 .243
8	−.037 .013 −.117 .037 .284
9	.029 .165 −.112 .127 .100 .280
10	SAMPLE SIZE: 167
11	LATENT VARIABLES: Thinking Feeling Acting
12	RELATIONSHIPS:
13	Tf = Thinking Feeling
14	Tc = Thinking
15	Fa = Feeling Acting
16	Fc = Feeling
17	At = Acting Thinking
18	Ac = Acting
19	PATH DIAGRAM
20	Number of Decimals = 3
21	END OF PROBLEM

Table A.7(a). SIMPLIS PATH DIAGRAM Output from an Analysis of the *HBI* Model in Figure 2.6

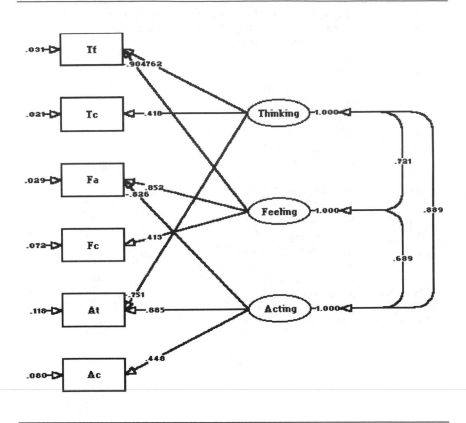

Table A.8. SIMPLIS Input File for the General Structural Equation Model in Figure 3.1, Example 3.1

1	!Example 3.1. A Structural Equation Model of Parents' on Respondent's SES
2	Observed Variables:
3	MoEd FaEd PaJntInc HSRank FinSucc ConCollg AcAbilty DriveAch SelfConf
4	DegreAsp ColContr Selctvty Degree OcPrestg Income
5	Correlation Matrix:
6	1
7	.610 1
8	.446 .531 1
9	.115 .128 .055 1
10	−.077 −.097 −.016 −.052 1
11	−.203 −.216 −.393 .002 −.018 1
12	.192 .216 .154 .493 −.086 −.079 1
13	−.042 −.017 −.023 .205 .063 .010 .251 1
14	.090 .112 .068 .269 −.021 −.043 .487 .327 1
15	.116 .122 .101 .194 −.008 .021 .236 .195 .206 1
16	.139 .205 .170 .049 −.125 .011 .119 .018 .056 .106 1
17	.255 .300 .293 .372 −.111 −.114 .382 .152 .216 .214 .294 1
18	.117 .129 .141 .189 −.025 −.067 .242 .184 .179 .253 .144 .254 1
19	.057 .084 .059 .153 −.002 .017 .163 .098 .090 .125 .110 .155 .481 1
20	.012 −.008 .093 .037 .157 −.060 .064 .096 .040 .025 −.020 .074 .106 .136 1
21	Standard Deviations:
22	1.229 1.511 2.649 .777 .847 .612 .744 .801 .782 1.014 .475 1.990 .962
23	1.591 1.627
24	Sample Size: 3094
25	Reorder Variables:
26	AcAbilty SelfConf DegreAsp Selctvty Degree OcPrestg MoEd FaEd PaJntInc HSRank
27	Latent Variables: AcMotiv ColgPres SES PaSES AcRank
28	Relationships:
29	AcAbilty = 1*AcMotiv
30	SelfConf DegreAsp = AcMotiv
31	Selctvty = 1*ColgPres
32	Degree = 1*SES
33	OcPrestg = SES
34	MoEd = 1*PaSES
35	FaEd PaJntInc = PaSES
36	HSRank = 1*AcRank
37	AcMotiv = PaSES AcRank
38	ColgPres = PaSES AcRank AcMotiv
39	SES = PaSES AcRank AcMotiv ColgPres
40	Set the Error Variance of HSRank to 0
41	Set the Error Variance of Selctvty to 0
42	Let the Errors between AcAbilty and SelfConf Correlate
43	Let the Errors between DegreAsp and Degree Correlate
44	Path Diagram
45	Number of Decimals = 3
46	LISREL Output: EF
47	End of Program

Table A.8(a). Partial SIMPLIS PATH DIAGRAM Output from an
Analysis of the Model in Figure 3.1: The Structural Portion

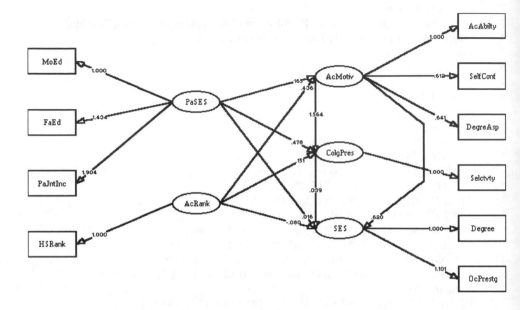

Table A.8(b). Partial SIMPLIS PATH DIAGRAM Output from an
Analysis of the Model in Figure 3.1: The Measurement Portions

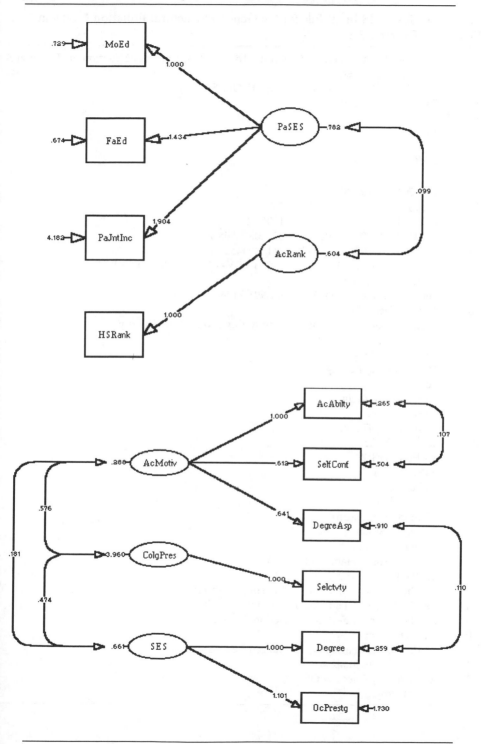

Table A.9. SIMPLIS Input File for the General Structural Equation Model in
Figure 3.5, Example 3.2

1	!Example 3.2. A Structural Equation Model of Sex, SES, and Situation on T, F, and A
2	Observed Variables:
3	Tf Tc Fa Fc At Ac Sex MoEd FaEd FaOcc Sit
4	Correlation Matrix:
5	1
6	.153 1
7	−.773 −.157 1
8	−.447 .579 .332 1
9	−.106 .054 −.320 .142 1
10	.083 .704 −.310 .487 .354 1
11	−.213 −.003 .086 .188 .136 .056 1
12	.042 .009 −.012 −.059 .036 .031 .052 1
13	−.041 .011 −.026 −.022 .061 .025 .081 .508 1
14	.054 .077 .052 .034 .056 .057 −.011 .363 .526 1
15	−.323 −.176 .495 .096 −.291 −.276 .004 −.046 −.020 −.083 1
16	Standard Deviations:
17	.660 .443 .684 .493 .533 .529 .500 1.991 2.059 1.578 .501
18	Sample Size: 167
19	Latent Variables: Thinking Feeling Acting BioSex SES Situatin
20	Relationships:
21	Tc = 1*Thinking
22	Tf = Thinking Feeling
23	Fc = 1*Feeling
24	Fa = Feeling Acting
25	Ac = 1*Acting
26	At = Acting Thinking
27	Sex = 1*BioSex
28	MoEd = 1*SES
29	FaEd = SES
30	FaOcc = SES
31	Sit = 1*Situatin
32	Thinking = Situatin
33	Feeling = Situatin
34	Acting = Situatin
35	Set the Error Variance of Sex to 0
36	Set the Error Variance of Sit to 0
37	Let the Errors of Thinking and Feeling Correlate
38	Let the Errors of Thinking and Acting Correlate
39	Let the Errors of Feeling and Acting Correlate
40	Path Diagram
41	Method of Estimation = Generalized Least Squares
42	Number of Decimals = 3
43	Admissibility Check = Off
44	End of Program

Table A.9(a). Partial SIMPLIS PATH DIAGRAM Output from an
Analysis of the Model in Figure 3.5: The Structural Portion

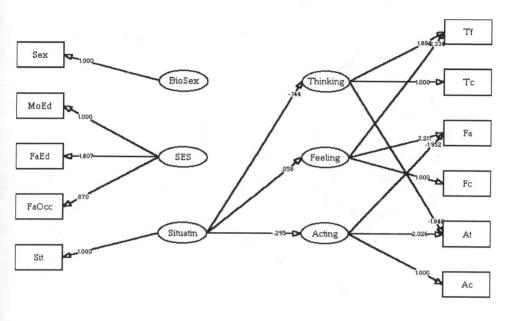

Table A.9(b). Partial SIMPLIS PATH DIAGRAM Output from an Analysis of the Model in Figure 3.5: The Measurement Portions

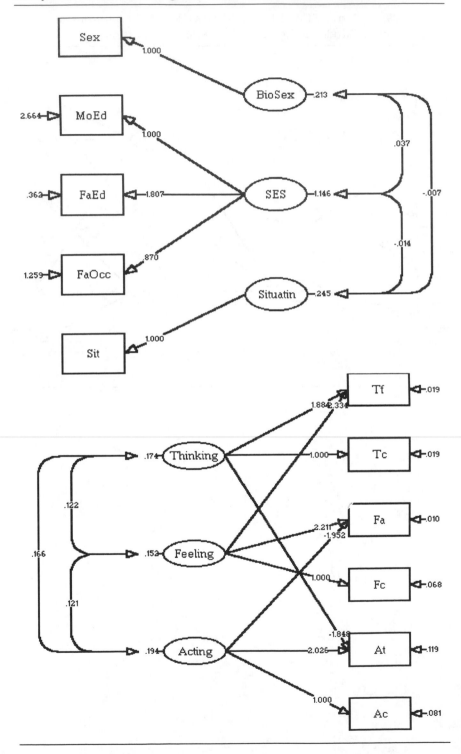

Location, Dispersion, and Association

Overview and Key Points

A meaningful study of structural equation modeling partially depends on a thorough understanding of some very fundamental statistical concepts. Clearly, not all pertinent issues can be reviewed within a short appendix such as this. However, as an introduction to some of the notation used throughout the book and a reminder of some basic statistical concepts, this appendix contains a brief review of the definitions and central properties of statistical expectation, variability, covariation, and standardization—all concepts of central importance to any area of applied statistics. Readers not familiar or comfortable with applying or interpreting the reviewed topics should consult appropriate sections within any of the recommended books listed at the end of this appendix. Specifically, the six key points briefly addressed in this appendix are as follows:

1. The expected value of a continuous variable can be viewed as the estimation of the value of a randomly selected score from the variable's distribution.
2. The mean of a distribution of scores from a continuous variable is used as a measure of the distribution's location. The mean is defined as the expected value of the variable.
3. The variance of a distribution of scores from a continuous variable is used as a measure of the distribution's dispersion. Variance is defined as the expected value of the squared deviations of the scores from their mean. The standard deviation of a distribution is the positive square root of the variance.
4. The covariance between two continuous variables is used as a measure of association between two variables. Covariance is the expected value of the products of deviations of the variables' scores from their respective means.

5. A standardized variable is a variable that has a distribution with a mean of 0 and a variance of 1. A continuous variable can be standardized by dividing each score's deviation from the distribution's mean by the distribution's standard deviation.
6. The Pearsonian correlation between two continuous variables can be viewed as the covariance between the corresponding two standardized variables.

Statistical Expectation

A Measure of a Distribution's Location

Given a distribution of N scores, X_k, $k = 1, \ldots, N$, of a variable X, the "best guess" at the value of X_k is defined as the *expected value* of X; formally,

$$E(X) = \sum_{k=1}^{N} X_k \mathrm{p}(X_k), \tag{B.1}$$

where $\mathrm{p}(X_k)$ is the probability of X_k being chosen, i.e., $\mathrm{p}(X_k) = f_k/N$ with f_k being the frequency of occurrence of the value X_k. If the values of the variable X are listed individually, $E(X)$ is one way to express the *location* of the distribution of the variable X. That is, using equation (B.1), the *mean* μ_X of X can be defined as

$$\mu_X = E(X) = \sum_{k=1}^{N} X_k(f_k/N) = \sum_{k=1}^{N} (X_k/N) = \frac{\sum_{k=1}^{N} X_k}{N}. \tag{B.2}$$

For example, suppose that variable X takes on the values $\{4, 3, 5, 8, 10\}$. Then, the mean of this set of scores is given by

$$\mu_X = E(X) = [4(1/5) + 3(1/5) + 5(1/5) + 8(1/5) + 10(1/5)]$$
$$= (4 + 3 + 5 + 8 + 10)/5 = 6.$$

A Measure of a Distribution's Dispersion

How far spread out are the values of the variable X in the distribution? Usually, the *variance* σ_X^2 of the variable X is used to measure the dispersion of scores and is defined as the mean squared deviation of scores from their mean, that is,

$$\sigma_X^2 = \mathrm{var}(X) = E([X - E(X)]^2) = E([X - \mu_X]^2) = \frac{\sum_{k=1}^{N} (X_k - \mu_X)^2}{N}, \tag{B.3}$$

where the numerator usually is referred to as the *sum-of-squares* (SS_X) associated with variable X.

Since the variance measures dispersion in squared units of the variable X, a related measure of dispersion is defined to enhance interpretability: *The standard deviation* of X, σ_X, is defined as the positive square root of the variance of X,

$$\sigma_X = \text{sd}(X) = \sqrt{\sigma_X^2} \tag{B.4}$$

and, thus, expresses score dispersion in the same units of measurement as the variable X.

For the above set of values of X, $\{4, 3, 5, 8, 10\}$, the variance and standard deviation can be computed as

$$\sigma_X^2 = [(4-6)^2 + (3-6)^2 + (5-6)^2 + (8-6)^2 + (10-6)^2]/5 = 6.8$$

and

$$\sigma_X = \sqrt{6.8} = 2.61.$$

A Measure of Association Between Two Variables

To numerically assess the direction and strength of the relationship or association between two continuous variables, say, X and Y, define the *covariance* σ_{XY} between X and Y as the expected value of the products of the deviations of the variables from their respected means, as in

$$\sigma_{XY} = \text{cov}(XY) = E([X - E(X)][Y - E(Y)]) = \frac{\sum_{k=1}^{N}(X_k - \mu_X)(Y_k - \mu_Y)}{N},$$

$$\tag{B.5}$$

where the numerator usually is referred to as the *cross-product* (CP_{XY}) associated with variables X and Y. For the variable X with values $\{4, 3, 5, 8, 10\}$ and mean $\mu_X = 6$, and the variable Y with values $\{0, 2, 6, 7, 10\}$ and mean $\mu_Y = 5$, the covariance between X and Y is

$$\sigma_{XY} = [(4-6)(0-5) + (3-6)(2-5) + (5-6)(6-5)$$

$$+ (8-6)(7-5) + (10-6)(10-5)]/5$$

$$= [10 + 9 + (-1) + 4 + 20]/5 = 8.4.$$

Five identities are very helpful when dealing with covariances and are used throughout the book (as an exercise, the reader is encouraged to use the above data to numerically verify these identities and then try to prove them mathematically). Consider variables X, Y, and Z, and let c be any constant. Then,

1. $\text{cov}(XY) = \text{cov}(YX)$; that is, a change in variable order does not change the value of the covariance between two variables;
2. $\text{cov}(cX) = 0$; a variable does not covary with a constant;
3. $\text{cov}(XX) = \text{var}(X)$; the covariance of a variable with itself is its variance;

4. $\text{cov}[(cX)Y] = (c)\text{cov}(XY)$; the multiplication of a variable by a constant c changes the variable's covariance with another variable by a factor of c; and, finally

5. $\text{cov}[X(Y + Z)] = \text{cov}(XY) + \text{cov}(XZ)$; that is, the covariance operator is distributive with respect to addition.

Now consider a variable Y that is a linear combination of another variable X; that is, $Y = c_0 + c_1 X_1$, where c_0 and c_1 are constants. Some algebraic manipulations using the definitions in equations (B.2), (B.3), and (B.5) and the identities just mentioned show that

$$E(Y) = c_0 + c_1 E(X_1) \tag{B.6}$$

and

$$\sigma_Y^2 = c_1^2 \sigma_{X_1}^2. \tag{B.7}$$

Thus, if a variable Y is a linear function of a variable X then its mean can be expressed as a linear function of the mean of X. In addition, its variance is a nonlinear function (with respect to the coefficient c_1) of the variance of X.

Similarly, if $Y = c_0 + c_1 X_1 + c_2 X_2$, i.e., a linear combination of two variables, X_1 and X_2, then its mean and variance are given by

$$E(Y) = c_0 + c_1 E(X_1) + c_2 E(X_2) \tag{B.8}$$

and

$$\sigma_Y^2 = c_1^2 \sigma_{X_1}^2 + c_2^2 \sigma_{X_2}^2 + 2c_1 c_2 \sigma_{X_1 X_2}. \tag{B.9}$$

For example, consider a variable X_1 with values $\{4, 3, 5, 8, 10\}$, mean $\mu_{X_1} = 6$, and $\sigma_{X_1}^2 = 6.8$, and X_2 with values $\{0, 2, 6, 7, 10\}$, $\mu_{X_2} = 5$, and $\sigma_{X_2}^2 = 12.8$. As was shown above, $\sigma_{X_1 X_2} = 8.4$. If, for example, $c_0 = 1$, $c_1 = 2$, and $c_2 = 3$, then the mean and variance of $Y = c_0 + c_1 X_1 + c_2 X_2 = 1 + (2)X_1 + (3)X_2$ are given by

$$E(Y) = 1 + 2(6) + 3(5) = 28$$

and

$$\sigma_Y^2 = 2^2(6.8) + 3^2(12.8) + 2(2)(3)(8.4) = 243.2.$$

In general, if the variable Y is expressed as a constant plus a linear combination of other variables, X_k, that is,

$$Y = c_0 + c_1 X_1 + c_2 X_2 + \cdots + c_{NX} X_{NX} = c_0 + \sum_{k=1}^{NX} c_k X_k, \tag{B.10}$$

where each c_k, $k = 0, 1, 2, \ldots, NX$, is a constant and NX is the total number of X variables, then the mean and variance of Y can be written as

$$E(Y) = c_0 + \sum_{k=1}^{NX} c_k E(X_k) \tag{B.11}$$

and

$$\sigma_Y^2 = \sum_{(\text{all } k, l)} c_k c_l \sigma_{X_k X_l}$$

$$= \sum_{(k=l)} c_k^2 \sigma_{X_k}^2 + \sum_{(k \neq l)} \sum c_k c_l \sigma_{X_k X_l}, \qquad (\text{B.12})$$

where $k, l = 1, 2, \ldots, NX$.

Statistical Standardization

Standardized Variables

Let X represent a variable with a given mean μ_X and variance σ_X^2. How can X be transformed into a variable Z_X with mean equal to 0 and variance equal to unity? Let $Z_X = c_0 + cX$, where c_0 and c are constants. Then, using equations (B.6) and (B.7),

$$E(Z_X) = c_0 + cE(X) = 0 \qquad (\text{B.13})$$

and

$$\sigma_{Z_X}^2 = c^2 \sigma_X^2 = 1. \qquad (\text{B.14})$$

Solving equations (B.13) and (B.14) for c and c_0 yields

$$c = 1/\sigma_X$$

and

$$c_0 = -E(X)/\sigma_X.$$

Thus, the *standardized variable* Z_X is given by

$$Z_X = c_0 + cX = (-E(X)/\sigma_X) + (1/\sigma_X)X = \frac{X - E(X)}{\sigma_X} = \frac{X - \mu_X}{\sigma_X}. \qquad (\text{B.15})$$

This transformed variable has a mean of 0 and a variance (and standard deviation) of 1. Similarly, given variable X, if a transformed variable D is to have a mean of 0 but an unchanged variance σ_X^2, then $D = [X - E(X)] = (X - \mu_X)$ is the appropriate transformation.

Again consider the variable X with values $\{4, 3, 5, 8, 10\}$, $\mu_X = 6$, and $\sigma_X = 2.61$. The set of standardized scores Z_X, computed using equation (B.15),

$$\{(4 - 6)/2.61, (3 - 6)/2.61, (5 - 6)/2.61, (8 - 6)/2.61, (10 - 6)/2.61\}$$

$$= \{-0.77, -1.15, -0.38, 0.77, 1.53\}$$

has a mean of 0 and a standard deviation of 1, as can be verified easily. Similarly, for the variable Y with values $\{0, 2, 6, 7, 10\}$, $\mu_Y = 5$, and $\sigma_Y = 3.58$, the set of associated standardized scores Z_Y is

$$\{-1.40, -0.84, 0.28, 0.56, 1.40\}.$$

A Standardized Measure of Association Between Two Variables

The computation of the covariance between two standardized continuous variables leads to the concept of Pearsonian correlation. Let X and Y be two unstandardized continuous variables with their corresponding standardized counterparts Z_X and Z_Y. Then,

$$\sigma_{Z_X Z_Y} = \text{cov}(Z_X Z_Y) = \text{cov}\left(\left[\frac{X - \mu_X}{\sigma_X}\right]\left[\frac{Y - \mu_Y}{\sigma_Y}\right]\right)$$

and, after some algebraic manipulations using the above covariance identities,

$$\sigma_{Z_X Z_Y} = \sigma_{XY}/\sigma_X \sigma_Y. \tag{B.16}$$

For example, using the definition formula of covariance in equation (B.5) to calculate the left side of equation (B.16) for variables Z_X and Z_Y in the above example leads to $\text{cov}(Z_X Z_Y) = 0.90$. This value equals the result of calculating the right side with $\sigma_{XY} = 8.4$, $\sigma_X = 2.61$, and $\sigma_Y = 3.58$.

The term on the right side of equation (B.16) is one way to define Pearson's *product-moment correlation coefficient* between two continuous variables X and Y, denoted here as ρ_{XY}; that is,

$$\rho_{XY} = \sigma_{XY}/\sigma_X \sigma_Y. \tag{B.17}$$

Recommended Readings

For a thorough introduction to concepts mentioned in this appendix, any elementary statistics text can be consulted. For the social scientist, books like the following might be particularly helpful:

Hays, W.L. (1988). *Statistics* (4th ed.). New York: Holt, Rinehart and Winston.
Hinkle, D.E., Wiersma, W., and Jurs, S.G. (1994). *Applied Statistics for the Behavioral Sciences* (3rd ed.). Boston: Houghton Mifflin.
Howell, D.C. (1992). *Statistical Methods for Psychology* (3rd ed.). Boston: PWS-Kent.
Keppel, G. (1991). *Design and Analysis: A Researcher's Handbook* (3rd ed.). Englewood Cliffs, NJ: Prentice-Hall.
Kirk, R.E. (1982). *Experimental Design* (2nd ed.). Belmont, CA: Brooks/Cole.
Marascuilo, L.A., and Serlin, R.C. (1988). *Statistical Methods for the Social and Behavioral Sciences*. New York: Freeman.

Matrix Algebra

Overview and Key Points

The mathematical foundations of many statistical techniques, including structural equation modeling, can be presented and discussed rather easily when using matrix formulations. In every textbook on elementary linear algebra and most books on intermediate applied statistics, matrix algebra is thoroughly discussed. Thus, it suffices here to review pertinent elementary definitions and properties that are used throughout the book. Specifically, in this appendix four key points are reviewed:

1. Matrix addition is an elementwise operation that is commutative, associative, and has an identity and inverse element.
2. Matrix multiplication is *not* an elementwise operation. It is *not* commutative in general, but it is associative and distributive with respect to matrix addition. An identity and, under certain conditions, an inverse element exists.
3. Determinants are unique numbers assigned to square matrices that are used throughout the more technical parts of this book.
4. The analysis of variance/covariance matrices of observed variables is at the center of structural equation modeling. Thus, an understanding of this type of matrix is of great importance.

Some Basic Definitions

A *matrix* is defined as a collection of numbers (called the *elements* of the matrix) organized by rows and columns. The *order* of a matrix gives the number of rows and columns. For example, the matrix **A**, given by

$$A = \begin{bmatrix} 5 & 17 \\ 6 & 3 \\ 0 & 11 \end{bmatrix},$$

is a matrix of order (3×2) since there are three rows and two columns. In general, if A is a $(p \times q)$ matrix, then A has p rows and q columns; the element that is in the ith row and the jth column of A is denoted by a_{ij}. If $p = q$, the A is said to be a *square matrix*. The *transpose* of a $(p \times q)$ matrix A, denoted by A', is a $(q \times p)$ matrix obtained by interchanging the rows and columns. Thus, for the above example,

$$A' = \begin{bmatrix} 5 & 6 & 0 \\ 17 & 3 & 11 \end{bmatrix}.$$

Note that $(A')' = A$. If $A = A'$, then A is called a *symmetric matrix*, and, if in addition, all off-diagonal elements are 0, then A is said to be a *diagonal matrix*. A $(p \times p)$ diagonal matrix with only ones on the diagonal is called the $(p \times p)$ *identity matrix*, denoted by I.

If a matrix A is symmetric and contains only zeros above or below the main diagonal, then A is called a *triangular matrix*. The *trace* of the matrix A, denoted by tr(A), is defined to be the sum of the diagonal elements in A. Finally, a $(p \times 1)$ matrix is called a *column vector*, while a $(1 \times q)$ matrix is called a *row vector*.

Algebra with Matrices

Matrix addition and *subtraction* are elementwise operations in the sense that adding or subtracting two matrices, A and B, results in a third matrix, $C = (A \pm B)$, whose elements are obtained by adding or subtracting corresponding elements in A and B. For example, if

$$A = \begin{bmatrix} 5 & 17 \\ 6 & 3 \\ 0 & 11 \end{bmatrix} \quad \text{and} \quad B = \begin{bmatrix} 0 & 3 \\ 10 & 9 \\ 5 & 12 \end{bmatrix},$$

then

$$C = A + B = \begin{bmatrix} 5 + 0 & 17 + 3 \\ 6 + 10 & 3 + 9 \\ 0 + 5 & 11 + 12 \end{bmatrix} = \begin{bmatrix} 5 & 20 \\ 16 & 12 \\ 5 & 23 \end{bmatrix} \quad \text{or}$$

$$C = A - B = \begin{bmatrix} 5 - 0 & 17 - 3 \\ 6 - 10 & 3 - 9 \\ 0 - 5 & 11 - 12 \end{bmatrix} = \begin{bmatrix} 5 & 14 \\ -4 & -6 \\ -5 & -1 \end{bmatrix}.$$

Note that only matrices of the same order can be added or subtracted.

If **A**, **B**, and **C** are all $(p \times q)$ matrices, then the following three properties are preserved under matrix addition:

1. the commutative law, i.e., $\mathbf{A} + \mathbf{B} = \mathbf{B} + \mathbf{A}$; and
2. the associative law, i.e., $(\mathbf{A} + \mathbf{B}) + \mathbf{C} = \mathbf{A} + (\mathbf{B} + \mathbf{C})$; furthermore,
3. $(\mathbf{A} + \mathbf{B})' = \mathbf{A}' + \mathbf{B}'$.

A $(p \times q)$ matrix, **0**, consisting only of zeros, serves as the *identity element* for matrix addition, i.e., $\mathbf{A} \pm \mathbf{0} = \mathbf{A}$. Finally, the *additive inverse* of **A**, denoted by $-\mathbf{A}$, is $(-1)\mathbf{A}$ [obtained by multiplying each element in **A** by the constant (-1)] with $\mathbf{A} + (-\mathbf{A}) = \mathbf{0}$.

Matrix multiplication, as opposed to matrix addition and subtraction, is *not* an elementwise operation. Instead, the $(p \times r)$ product, **AB**, of a $(p \times q)$ matrix **A** with a $(q \times r)$ matrix **B** is defined as follows: let a_{ik} and b_{kj} denote elements in **A** and **B**, respectively. Then, the elements $(ab)_{ij}$ of **AB** are defined by

$$(ab)_{ij} = \sum_{k=1}^{q} a_{ik}b_{kj}, \quad \text{for } i = 1, 2, \ldots, p \quad \text{and} \quad j = 1, 2, \ldots, r.$$

Note that the number of columns in **A** must be equal to the number of rows in **B** for the product **AB** to be defined. Consider the following example: Let

$$\mathbf{A} = \begin{bmatrix} 5 & 7 \\ 6 & 3 \\ 0 & 1 \end{bmatrix} \quad \text{and} \quad \mathbf{B} = \begin{bmatrix} 10 & 9 \\ 5 & 2 \end{bmatrix};$$

then

$$\mathbf{AB} = \begin{bmatrix} (5)(10) + (7)(5) & (5)(9) + (7)(2) \\ (6)(10) + (3)(5) & (6)(9) + (3)(2) \\ (0)(10) + (1)(5) & (0)(9) + (1)(2) \end{bmatrix} = \begin{bmatrix} 85 & 59 \\ 75 & 60 \\ 5 & 2 \end{bmatrix}.$$

In general, if **A**, **B**, and **C** are matrices of the appropriate orders, then the following three properties are preserved under matrix multiplication:

1. the associative law, i.e., $(\mathbf{AB})\mathbf{C} = \mathbf{A}(\mathbf{BC})$; and
2. the distributive law with respect to matrix addition, i.e., $\mathbf{A}(\mathbf{B} + \mathbf{C}) = \mathbf{AB} + \mathbf{AC}$ and $(\mathbf{A} + \mathbf{B})\mathbf{C} = \mathbf{AC} + \mathbf{BC}$; furthermore,
3. $(\mathbf{AB})' = \mathbf{B}'\mathbf{A}'$.

Note, however, that matrix multiplication is *not* commutative, that is, in general, **AB** is not equal to **BA**. The identity matrix **I** serves as the identity element for matrix multiplication:

$$\mathbf{AI} = \mathbf{IA} = \mathbf{A}.$$

The *multiplicative inverse* of a matrix **A**, denoted by $\mathbf{A}^{-1}$, does not always exist; if it does, then **A** is called *nonsingular* or *invertible*; otherwise, **A** is called *singular*. If $\mathbf{A}$ is invertible, then $\mathbf{AA}^{-1} = \mathbf{A}^{-1}\mathbf{A} = \mathbf{I}$ (there are a variety of algorithms available to calculate a matrix inverse—if it exists; consult any elementary textbook on linear algebra). If $\mathbf{A}$ and $\mathbf{B}$ are both invertible matrices of orders $(p \times q)$ and $(q \times r)$, respectively, then

1. $(\mathbf{A}^{-1})^{-1} = \mathbf{A}$; and
2. $(\mathbf{AB})^{-1} = \mathbf{B}^{-1}\mathbf{A}^{-1}$; furthermore,
3. $(\mathbf{A}^{-1})' = (\mathbf{A}')^{-1}$.

Finally, the concept of the *determinant* of a square matrix $\mathbf{A}$, denoted by $\det(\mathbf{A})$ or $|\mathbf{A}|$, is important in statistics. Loosely defined, $|\mathbf{A}|$ is a unique number assigned to $\mathbf{A}$ that must satisfy certain properties. Depending on the order of $\mathbf{A}$, $|\mathbf{A}|$ is calculated by a certain algorithm. For example, the determinant of a (2×2) matrix is calculated as follows: If

$$\mathbf{A} = \begin{bmatrix} a & b \\ c & d \end{bmatrix},$$

then $|\mathbf{A}|$ is defined as $|\mathbf{A}| = ad - bc$. Thus, if

$$\mathbf{A} = \begin{bmatrix} 2 & 5 \\ 1 & 3 \end{bmatrix},$$

then $|\mathbf{A}| = (2)(3) - (5)(1) = 1$.

In general, if $\mathbf{A}$ and $\mathbf{B}$ are two matrices of the appropriate orders, then the following five properties of determinants hold:

1. $|\mathbf{A} \pm \mathbf{B}| = |\mathbf{A}| \pm |\mathbf{B}|$;
2. $|\mathbf{AB}| = |\mathbf{A}||\mathbf{B}|$;
3. $|\mathbf{A}^{-1}| = 1/|\mathbf{A}|$, provided $\mathbf{A}$ is nonsingular;
4. if $|\mathbf{A}| = 0$, then $\mathbf{A}$ is singular; otherwise, $\mathbf{A}$ is invertible; furthermore,
5. $|\mathbf{A}'| = |\mathbf{A}|$.

The Variance/Covariance Matrix

A central concept underlying structural equation modeling is the analysis of a variance/covariance matrix based on data from N individuals on NX observed variables, $X_1, \ldots, X_{NX}$. Define the $(N \times NX)$ *data matrix* $\mathbf{X}$ as the matrix of deviation scores from variable means of the N individuals on NX observed variables. First, note that $\mathbf{X}'\mathbf{X}$ is the $(NX \times NX)$ matrix that has the sum-of-squares $(SS_i = \sum_{k=1}^{N} (X_{ik} - E(X_i))^2)$ of the NX variables X_i as its diagonal elements and the cross-products $(CP_{ij(i \neq j)} = \sum_{k=1}^{N} (X_{ik} - E(X_i)) \times (X_{jk} - E(X_j)))$ of the variables as its off-diagonal elements (also see Appendix B for the definitions of sum-of-squares and cross-products). Second, the expected value $E(\mathbf{A})$ of a matrix $\mathbf{A}$ containing variables as its elements is the matrix containing the expected values of each of the elements of $\mathbf{A}$; that is, the expected value operator is an elementwise operator with respect to matrices (see Appendix B for a definition of the expected value of a variable). Now, it follows that the matrix

$$E(\mathbf{X'X}) = E\left(\begin{bmatrix} SS_1 & CP_{12} & \cdots & CP_{1NX} \\ CP_{21} & SS_2 & \cdots & CP_{2NX} \\ \vdots & \vdots & \ddots & \vdots \\ CP_{NX1} & CP_{NX2} & \cdots & SS_{NX} \end{bmatrix}\right)$$

$$= \begin{bmatrix} SS_1/N & CP_{12}/N & \cdots & CP_{1NX}/N \\ CP_{21}/N & SS_2/N & \cdots & CP_{2NX}/N \\ \vdots & \vdots & \ddots & \vdots \\ CP_{NX1}/N & CP_{NX2}/N & \cdots & SS_{NX}/N \end{bmatrix} = \begin{bmatrix} \sigma_1^2 & \sigma_{12} & \cdots & \sigma_{1NX} \\ \sigma_{21} & \sigma_2^2 & \cdots & \sigma_{2NX} \\ \vdots & \vdots & \ddots & \vdots \\ \sigma_{NX1} & \sigma_{NX2} & \cdots & \sigma_{NX}^2 \end{bmatrix},$$

called the *variance/covariance matrix* Σ of the NX observed variables, contains the variances of the NX variables on its diagonal and the covariances between the variables as its off-diagonal elements. When all variables are standardized, the variance/covariance matrix Σ becomes a *correlation matrix* with ones on its diagonal and the Pearsonean correlations between variables as its off-diagonal elements.

Consider an example: Suppose three individuals obtain scores of $\{2, 4, 6\}$ on some variable X_1 and scores $\{8, 2, 5\}$ on another variable, say, X_2. Clearly, $\mu_{X_1} = E(X_1) = 4$ and $\mu_{X_2} = E(X_2) = 5$ (see Appendix B). Then the (3×2) data matrix $\mathbf{X}$ consisting of deviation scores from the means is given by

$$\mathbf{X} = \begin{bmatrix} -2 & 3 \\ 0 & -3 \\ 2 & 0 \end{bmatrix}.$$

Now,

$$\mathbf{X'X} = \begin{bmatrix} -2 & 0 & 2 \\ 3 & -3 & 0 \end{bmatrix}\begin{bmatrix} -2 & 3 \\ 0 & -3 \\ 2 & 0 \end{bmatrix} = \begin{bmatrix} 8 & -6 \\ -6 & 18 \end{bmatrix} = \begin{bmatrix} SS_1 & CP_{12} \\ CP_{21} & SS_2 \end{bmatrix}$$

and

$$E(\mathbf{X'X})$$

$$= (1/3)\begin{bmatrix} 8 & -6 \\ -6 & 18 \end{bmatrix} = \begin{bmatrix} 8/3 & -6/3 \\ -6/3 & 18/3 \end{bmatrix} = \begin{bmatrix} 2.67 & -2.00 \\ -2.00 & 6.00 \end{bmatrix} = \begin{bmatrix} \sigma_{X_1}^2 & \sigma_{X_1 X_2} \\ \sigma_{X_2 X_1} & \sigma_{X_2}^2 \end{bmatrix}$$

is the variance/covariance matrix with $\sigma_{X_1 X_2} = \text{cov}(X_1 X_2) = -2.00$, $\sigma_{X_1}^2 = 2.67$, and $\sigma_{X_2}^2 = 6.00$, as can be verified easily by using the formulae presented in Appendix B.

Recommended Readings

For a thorough introduction to the topics reviewed in this appendix, any basic text on linear algebra can be consulted. See, for example,

Anton, H. (1991). *Elementary Linear Algebra* (6th ed.). New York: John Wiley & Sons.

Kolman, B. (1993). *Introductory Linear Algebra with Applications* (5th ed.). New York: Macmillan.

Very useful references for statisticians are the two listed below. Both deal exclusively with statistics-related matrix algebra; the latter is introductory while the former is a more advanced text in differential matrix calculus.

Magnus, J.R., and Neudecker (1988). *Matrix Differential Calculus with Applications in Statistics and Econometrics.* New York: John Wiley & Sons.
Searle, S.R. (1982). *Matrix Algebra Useful for Statistics.* New York: John Wiley & Sons.

Finally, some applied multivariate statistics texts have good summaries of fundamentals of matrix algebra. In particular, see

Stevens, J. (1992). *Applied Multivariate Statistics for the Social Sciences* (2nd ed.). Hillsdale, NJ: Lawrence Erlbaum.
Tatsuoka, M.M. (1988). *Multivariate Analysis* (2nd ed.). New York: Macmillan.

Descriptive Statistics for the SES Analysis

Table D.1. Coding Schema for Variables in the SES Analysis

Variable	Code	Variable	Code	Variable	Code
1. Mother's Educational level (*MoEd*);	1 = grammar school or less 2 = some high school	6. Concern About Financing College (*ConCollg*)	1 = no concern 2 = some concern 3 = major concern	12. College Selectivity (*Selctvty*); average SAT	1 = less than 775 2 = 775 to 849 3 = 850 to 924
2. Father's Educational level (*FaEd*)	3 = high school graduate 4 = some college 5 = college degree 6 = postgraduate degree	7. Academic Ability (*AcAbilty*); 8. Drive to Achieve (*DriveAch*); 9. Self-Confidence (*SelfConf*); all are self-ratings	1 = lowest 10% 2 = below average 3 = average 4 = above average 5 = highest 10%		4 = 925 to 999 5 = 1,000 to 1,074 6 = 1,075 to 1,149 7 = 1,150 to 1,224 8 = 1,225 to 1,299 9 = 1,300 or more
3. Parents' Joint Income (*PaJntInc*)	1 = less than $ 4,000 2 = $ 4,000 to $ 5,999 3 = $ 6,000 to $ 7,999 4 = $ 8,000 to $ 9,999 5 = $10,000 to $12,499 6 = $12,500 to $14,999 7 = $15,000 to $19,999 8 = $20,000 to $24,999 9 = $25,000 to $29,999 10 = $30,000 to $34,999 11 = $35,000 to $39,999 12 = $40,000 or more			13. Highest Held Academic Degree (*Degree*)	1 = high school diploma (or equivalent) 2 = vocational certificate 3 = associate 4 = bachelor's 5 = master's 6 = doctorate

Table D.1 (*cont.*)

Variable	Code	Variable	Code	Variable	Code
4. High School Rank (*HSRank*)	1 = fourth quarter 2 = third quarter 3 = second quarter 4 = top quarter	10. Degree Aspiration (*DegreAsp*)	1 = none 2 = associate 3 = bachelor's 4 = master's 5 = doctorate	14. Occupational Prestige (*OcPrestg*)	SES (Duncan, 1961); rescaled by a factor of 0.1 (see Bentler, 1993, p. 20)
5. Want to be Financially Successful (*FinSucc*)	1 = not important 2 = somewhat important 3 = very important 4 = essential	11. College Control (*ColContr*)	1 = public 2 = private	15. Income 5 years after graduation (*Income*)	1 = none 2 = less than $ 7,000 3 = $ 7,000 to $ 9,999 4 = $10,000 to $14,999 5 = $15,000 to $19,999 6 = $20,000 to $24,999 7 = $25,000 to $29,999 8 = $30,000 to $34,999 9 = $35,000 to $39,999 10 = $40,000 or more

Note: Information in Table D.1 is taken from Mueller (1988) with permission from the publisher.

Table D.2. Means, Standard Deviations, and Correlations for the Male Subsample ($n = 3094$ based on listwise deletion)

Variables	μ	σ	1	2	3	4	5	6	7	8	9	10	11	12	13	14
1. MoEd	3.567	1.229														
2. FaEd	3.747	1.511	.610													
3. PaIntInc	5.884	2.649	.446	.531												
4. HSRank	3.420	.777	.115	.128	.055											
5. FinSucc	2.379	.847	−.077	−.097	−.016	−.052										
6. ConCollg	1.784	.612	−.203	−.216	−.393	.002	−.018									
7. AcAbilty	3.902	.744	.192	.216	.154	.493	−.086	−.079								
8. DriveAch	3.734	.801	−.042	−.017	−.023	.205	.063	.010	.251							
9. SelfConf	3.563	.782	.090	.112	.068	.269	−.021	−.043	.487	.327						
10. DegreAsp	4.003	1.014	.116	.122	.101	.194	−.008	.021	.236	.195	.206					
11. ColContr	1.655	.475	.139	.205	.170	.049	−.125	.011	.119	.018	.056	.106				
12. Selctvty	5.016	1.990	.255	.300	.293	.372	−.111	−.114	.382	.152	.216	.214	.294			
13. Degree	4.535	.962	.117	.129	.141	.189	−.025	−.067	.242	.184	.179	.253	.144	.254		
14. OcPrestg	6.184	1.591	.057	.084	.059	.153	−.002	.017	.163	.098	.090	.125	.110	.155	.481	
15. Income	4.756	1.627	.012	−.008	.093	.037	.157	−.060	.064	.096	.040	.025	−.020	.074	.106	.136

Table D.3. Means, Standard Deviations, and Correlations for the Female Subsample ($n = 3833$ based on listwise deletion)

Variables	μ	σ	1	2	3	4	5	6	7	8	9	10	11	12	13	14
1. MoEd	3.712	1.254														
2. FaEd	3.858	1.526	.605													
3. PaJntInc	5.792	2.619	.418	.522												
4. HSRank	3.589	.681	.092	.104	.082											
5. FinSucc	2.039	.782	−.063	−.088	−.041	−.094										
6. ConCollg	1.858	.620	−.233	−.267	−.408	.009	−.023									
7. AcAbility	3.855	.710	.181	.204	.194	.440	−.096	−.039								
8. DriveAch	3.787	.753	−.004	−.005	−.002	.180	.065	.040	.278							
9. SelfConf	3.376	.763	.101	.095	.078	.217	−.017	−.011	.490	.334						
10. DegreAsp	3.672	.900	.074	.053	.036	.116	−.002	.053	.188	.208	.186					
11. ColContr	1.684	.465	.158	.203	.157	.066	−.091	−.023	.126	.043	.076	.090				
12. Selctvty	4.653	1.836	.264	.327	.285	.289	−.132	−.073	.355	.092	.153	.201	.256			
13. Degree	4.325	.687	.073	.089	.066	.092	−.006	−.010	.154	.152	.140	.232	.088	.196		
14. OcPrestg	6.132	1.303	.003	−.005	.004	.037	.007	−.020	.065	.084	.075	.123	−.015	.032	.335	
15. Income	3.816	1.371	−.001	.019	.040	.061	.090	−.015	.080	.115	.070	.090	.022	.107	.226	.168

Note: Data in Tables D.2 and D.3 are taken from Mueller (1988) with permission from the publisher.

Descriptive Statistics for the HBI Analysis

Table E.1. Coding Schema for Variables in the HBI Analysis

Variable	Code	Variable	Code	Variable	Code
HBI Scales 1. *Tf** 2. *Tc* 3. *Fa** 4. *Fc* 5. *At** 6. *Ac*	see the *HBI* manual, Hutchins & Mueller (1992)	8. Mother's Education (*MoEd*); 9. Father's Education (*FaEd*)	1 = less than high school 2 = high school graduate 3 = less than 2 years of vocational, trade, or business school 4 = two years or more of vocational, trade, or business school or less than 2 years of college 5 = two years or more of college 6 = finished college 7 = Master's degree or equivalent 8 = PhD, MD, or other advanced degree	11. Situation specificity (*Situatin*)	0 = "How do I view myself as a student?" 1 = "How do I view myself when confronted with a close friend in emotional distress?"

Table E.1 (*cont.*)

Variable	Code	Variable	Code	Variable	Code
7. *Sex*	0 = male 1 = female	10. Father's occupation (*FaOcc*)	SES (Duncan, 1961)*		

*Rescaled by a factor of 0.1 (see Bentler, 1993, p. 20)

Table E.2. Means, Standard Deviations, and Correlations for the *HBI* Analysis ($n = 167$ based on pairwise deletion)

Variables	$\hat{\mu}$	$\hat{\sigma}$	1	2	3	4	5	6	7	8	9	10
1. *Tf*	1.09	.660										
2. *Tc*	2.01	.443	.153									
3. *Fa*	1.51	.684	-.773	-.157								
4. *Fc*	2.13	.493	-.447	.579	.332							
5. *At*	1.07	.533	-.106	.054	-.320	.142						
6. *Ac*	1.87	.529	.083	.704	-.310	.487	.354					
7. *Sex*	.46	.500	-.213	-.003	.086	.188	.136	.056				
8. *MoEd*	4.37	1.991	.042	.009	-.012	-.059	.036	.031	.052			
9. *FaEd*	5.50	2.059	-.041	.011	-.026	-.022	.061	.025	.081	.508		
10. *FaOcc*	6.06	1.578	.054	.077	.052	.034	.056	.057	-.011	.363	.526	
11. *Situatin*	.49	.501	-.323	-.176	.495	.096	-.291	-.276	.004	-.046	-.020	-.083

Note: Data in Tables E.1 and E.2 are taken from Mueller (1987) with permission from the author.

References

Aiken, L.S., and West, S.G. (1991). *Multiple Regression: Testing and Interpreting Interactions*. Newbury Park, CA: Sage.

Alwin, D.F., and Hauser, R.M. (1975). The decomposition of effects in path analysis. *American Sociological Review, 40*, 37–47.

Allen, M.J., and Yen, W.M. (1979). *Introduction to Measurement Theory*. Belmont, CA: Wadsworth.

Anderson, J.G., and Evans, F.B. (1974). Causal models in educational research: Recursive models. *American Educational Research Journal, 11*, 29–39.

Anton, H. (1991). *Elementary Linear Algebra* (6th ed.). New York: John Wiley & Sons.

Asher, H.B. (1983). *Causal Modeling* (2nd ed.). Newbury Park, CA: Sage.

Baumrind, D. (1983). Specious causal attributions in the social sciences. *Journal of Personality and Social Psychology, 45*, 1289–1298.

Bentler, P.M. (1980). Multivariate analysis with latent variables: Causal modeling. *Annual Review of Psychology, 31*, 419–456.

Bentler, P.M. (1986). Structural modeling and Psychometrika: An historical perspective on growth and achievements. *Psychometrika, 51*(1), 35–51.

Bentler, P.M. (1990). Comparative fit indexes in structural models. *Psychological Bulletin, 107*(2), 238–246.

Bentler, P.M. (1993). *EQS: Structural Equations Program Manual*. Los Angeles: BMDP Statistical Software.

Bentler, P.M., and Bonett, D.G. (1980). Significance tests and goodness of fit in the analysis of covariance structures. *Psychological Bulletin, 88*, 588–606.

Bentler, P.M., and Weeks, D.G. (1979). Interrelations among models for the analysis of moment structures. *Multivariate Behavioral Research, 14*, 169–185.

Bentler, P.M., and Weeks, D.G. (1980). Linear structural equations with latent variables. *Psychometrika, 45*, 289–308.

Bentler, P.M., and Wu, E.J.C. (1993). *EQS/Windows: User's Guide*. Los Angeles: BMDP Statistical Software.

Berry, W.D. (1984). *Nonrecursive Causal Models*. Newbury Park, CA: Sage.

Blalock, H.M. (1964). *Causal Inferences in Nonexperimental Research*. Chapel Hill: The University of North Carolina Press.

Blalock, H.M. (Ed.). (1985a). *Causal Models in the Social Sciences* (2nd ed.). New York: Aldine.

Blalock, H.M. (Ed.). (1985b). *Causal Models in Panel and Experimental Designs.* New York: Aldine.

Bollen, K.A. (1987). Total, direct, and indirect effects in structural equation models. In C. Clogg (Ed.), *Sociological Methodology 1987.* San Francisco: Jossey Bass.

Bollen, K.A. (1989). *Structural Equations with Latent Variables.* New York: John Wiley & Sons.

Bollen, K.A., and Long, J.S. (Eds.). (1993). *Testing Structural Equation Models.* Newbury Park, CA: Sage.

Browne, M.W. (1974). Generalized least squares estimators in the analysis of covariance structures. *South African Statistical Journal, 8,* 1–24.

Browne, M.W. (1982). Covariance structures. In D.M. Hawkins (Ed.), *Topics in Applied Multivariate Analysis* (pp. 72–141). London: Cambridge University Press.

Browne, M.W. (1984). Asymptotically distribution-free methods for the analysis of covariance structures. *British Journal of Mathematical and Statistical Psychology, 37,* 62–83.

Browne, M.W., and Cudeck, R. (1989). Single sample cross-validation indices for covariance structures. *Multivariate Behavioral Research, 24,* 445–455.

Byrne, B.M. (1989). *A Primer of LISREL: Basic Applications and Programming for Confirmatory Factor Analytic Models.* New York: Springer-Verlag.

Byrne, B.M. (1994). *Structural Equation Modeling with EQS and EQS/Windows: Basic Concepts, Applications, and Programming.* Thousand Oaks, CA: Sage.

Cliff, N. (1983). Some cautions concerning the application of causal modeling methods. *Multivariate Behavioral Research, 18,* 115–126.

Cohen, J., and Cohen, P. (1983). *Applied Multiple Regression/Correlation Analysis for the Behavioral Sciences* (2nd ed.). Hillsdale, NJ: Lawrence Erlbaum.

Crocker, L., and Algina, J. (1986). *Introduction to Classical and Modern Test Theory.* Orlando, FL: Holt, Rinehart and Winston.

Cronbach, L.J. (1951). Coefficient alpha and the internal structure of tests. *Psychometrika, 16,* 297–334.

Cudeck, R., and Browne, M.W. (1983). Cross-validation of covariance structures. *Multivariate Behavioral Research, 18,* 147–167.

Davis, J.A. (1985). *The Logic of Causal Order.* Newbury Park, CA: Sage.

Draper, N.R., and Smith, H. (1981). *Applied Regression Analysis* (2nd ed.). New York: John Wiley & Sons.

Duncan, O.D. (1961). Properties and characteristics of the socioeconomic index. In A.J. Reiss (Ed.), *Occupations and Social Status* (pp. 139–161). New York: Free Press.

Duncan, O.D. (1966). Path analysis: Sociological examples. *American Journal of Sociology, 72,* 1–16.

Duncan, O.D. (1975). *Introduction to Structural Equation Models.* New York: Academic Press.

Duncan, O.D., Haller, A.O., and Portes, A. (1968). Peer influence on aspiration: A reinterpretation. *American Journal of Sociology, 74,* 119–134.

Fox, J. (1980). Effect analysis in structural equation models. *Sociological Methods and Research, 9,* 3–28.

Freedman, D.A. (1987). As others see us: A case study in path analysis. *Journal of Educational Statistics, 12*(2), 101–128.

Glymour, C., Scheines, R., Spirtes, P., and Kelly, K. (1987). *Discovering Causal Structure: Artificial Intelligence, Philosophy of Science, and Statistical Modeling.* Orlando, FL: Academic Press.

Goldberger, A.S. (1964). *Econometric Theory.* New York: John Wiley & Sons.

Goldberger, A.S. (1971). Econometrics and psychometrics: A survey of comunalities. *Psychometrika, 36,* 83–107.

Gorsuch, R.L. (1983). *Factor Analysis* (2nd ed.). Hillsdale, NJ: Lawrence Erlbaum.

Hayduk, L.A. (1987). *Structural Equation Modeling with LISREL*. Baltimore: Johns Hopkins University Press.

Hays, W.L. (1988). *Statistics* (4th ed.). New York: Holt, Rinehart and Winston.

Hinkle, D.E., Wiersma, W., and Jurs, S.G. (1994). *Applied Statistics for the Behavioral Sciences* (3rd ed.). Boston: Houghton Mifflin.

Holland, P.W. (1986). Statistics and causal inference. *Journal of the American Statistical Association, 81*(396), 945–960.

Howell, D.C. (1992). *Statistical Methods for Psychology* (3rd ed.). Boston: PWS-Kent.

Huberty, C.J., and Wisenbaker, J.M. (1992). Variable importance in multivariate group comparisons. *Journal of Educational Statistics, 17*, 75–91.

Hutchins, D.E. (1979). Systematic counseling: The T-F-A model for counselor intervention. *The Personnel and Guidance Journal, 57*, 529–531.

Hutchins, D.E. (1982). Ranking major counseling strategies with the T-F-A/matrix system. *The Personnel and Guidance Journal, 60*, 427–431.

Hutchins, D.E. (1984). Improving the couseling relationship. *The Personnel and Guidance Journal, 62*, 572–575.

Hutchins, D.E. (1992). *The Hutchins Behavior Inventory*. Palo Alto, CA: Consulting Psychologists Press.

Hutchins, D.E., and Cole, C.G. (1992). *Helping Relationships and Strategies* (2nd ed.). Pacific Grove, CA: Brooks/Cole.

Hutchins, D.E., and Mueller, R.O. (1992). *Manual for the Hutchins Behavior Inventory*. Palo Alto, CA: Consulting Psychologists Press.

James, L.R., Mulaik, S.A., and Brett, J. (1982). *Causal Analysis: Models, Assumptions, and Data*. Beverly Hills, CA: Sage.

Jöreskog, K.G. (1967). Some contributions to maximum likelihood factor analysis. *Psychometrika, 32*, 443–482.

Jöreskog, K.G. (1969). A general approach to confirmatory maximum likelihood factor analysis. *Psychometrika, 34*, 183–202.

Jöreskog, K.G. (1970). A general method for analysis of covariance structures. *Biometrika, 57*, 239–251.

Jöreskog, K.G. (1973). A general method for estimating a linear structural equation system. In A.S. Goldberger and O.D. Duncan (Eds.), *Structural Equation Models in the Social Siences* (pp. 85–112). New York: Seminar Press.

Jöreskog, K.G. (1977). Structural equation models in the social sciences: Specification, estimation and testing. In P.R. Krishnaiah (Ed.), *Application of Statistics* (pp. 265–287). Amsterdam: North-Holland.

Jöreskog, K.G., and Goldberger, A.S. (1972). Factor analysis by generalized least squares. *Psychometrika, 37*, 243–250.

Jöreskog, K.G., and Sörbom, D. (1981). Analysis of linear structural relationships by maximum likelihood and least squares methods. Research Report 81-8, University of Uppsala, Sweden.

Jöreskog, K.G., and Sörbom, D. (1993a). *LISREL 8 User's Reference Guide*. Chicago: Scientific Software International.

Jöreskog, K.G., and Sörbom, D. (1993b). *LISREL 8: Structrual Equation Modeling with the SIMPLIS Command Language*. Chicago: Scientific Software.

Jöreskog, K.G., and Sörbom, D. (1993c). *PRELIS 2: A Program for Multivariate Data Screening and Data Summarization: A Preprocessor for LISREL*. Chicago: Scientific Software International.

Jöreskog, K.G., and Van Thillo, M. (1973). LISREL—A general computer program for estimating a linear structural equation system involving multiple indicators of unmeasured variables. Research Report 73-5, Department of Statistics, Uppsala University, Sweden.

Kaplan, D. (1990). Evaluating and modifying covariance structure models: A review and recommendation. *Multivariate Behavioral Research, 25*(2), 137–155.

Keesling, J.W. (1972). Maximum likelihood approaches to causal analysis. Unpublished doctoral dissertation, University of Chicago, Chicago, IL.

Kenny, D.A. (1979). *Correlation and Causation*. New York: John Wiley & Sons.

Keppel, G. (1991). *Design and Analysis: A Researcher's Handbook* (3rd ed.). Englewood Cliffs, NJ: Prentice-Hall.

Kirk, R.E. (1982). *Experimental Design* (2nd ed.). Belmont, CA: Brooks/Cole.

Kleinbaum, D.G., Kupper, L.L., and Muller, K.E. (1988). *Applied Regression Analysis and Other Multivariable Methods* (2nd ed.). Boston: PWS-Kent.

Kolman, B. (1993). *Introductory Linear Algebra with Applications* (5th ed.). New York: Macmillan.

Lawley, D.N. (1940). The estimation of factor loadings by the method of maximum likelihood. *Proceedings of the Royal Society of Edinburgh, A60*, 64–82.

Lawley, D.N. (1967). Some new results in maximum likelihood factor analysis. *Proceedings of the Royal Society of Edinburgh, A67*, 256–264.

Ling, R. (1983). Review of "Correlation and Causality" by David Kenny. *Journal of the American Statistical Association, 77*, 489–491.

Loehlin, J.C. (1992). *Latent Variable Models: An Introduction to Factor, Path, and Structural Analysis* (2nd ed.). Hillsdale, NJ: Lawrence Erlbaum.

Long, J.S. (1983a). *Confirmatory Factor Analysis*. Beverly Hills, CA: Sage.

Long, J.S. (1983b). *Covariance Structure Models: An Introduction to LISREL*. Beverly Hills, CA: Sage.

Lord, F.M., and Novick, M.R. (1968). *Statistical Theories of Mental Test Scores*. Menlo Park, CA: Addison-Wesley.

Magnus, J.R., and Neudecker, H. (1988). *Matrix Differential Calculus with Applications in Statistics and Econometrics*. New York: John Wiley & Sons.

Maiti, S.S., and Mukherjee, B.N. (1990). A note on the distributional properties of the Jöreskog-Sörbom fit indices. *Psychometrika, 55*, 721–726.

Marascuilo, L.A., and Serlin, R.C. (1988). *Statistical Methods for the Social and Behavioral Sciences*. New York: Freeman.

Marini, M.M., and Singer, B. (1988). Causality in the social sciences. In C.C. Clogg (Ed.), *Sociological Methodology: Vol. 18* (pp. 347–409). Washington, DC: American Sociological Association.

Marsh, H.W., Balla, J.R., and McDonald, R.P. (1988). Goodness-of-fit indexes in confirmatory factor analysis: The effect of sample size. *Psychological Bulletin, 103*(3), 391–410.

McDonald, R.P. (1978). A simple comprehensive model for the analysis of covariance structures. *British Journal of Mathematical and Statistical Psychology, 31*, 59–72.

McDonald, R.P. (1980). A simple comprehensive model for the analysis of covariance structures: Some remarks on applications. *British Journal of Mathematical and Statistical Psychology, 33*, 161–183.

McDonald, R.P. (1985). *Factor Analysis and Related Methods*. Hillsdale, NJ: Lawrence Erlbaum.

Mueller, R.O. (1987). The effects of gender, socioeconomic status, and situation specificity on thinking, feeling, and acting. *Dissertation Abstracts International, 48*, 1441A–1442A. (University Micofilms No. 87-19, 038)

Mueller, R.O. (1988). The impact of college selectivity on income for men and women. *Research in Higher Education, 29*(2), 175–191.

Mueller, R.O., and Dupuy, P.J. (1992, October). Building and interpreting causal models: The role of causality. Paper presented at the meeting of the Mid-Western Educational Research Association, Chicago, IL.

Mueller, R.O., and Hutchins, D.E. (1991). Behavior orientations of student residence hall counselors: An application of the TFA system. *The Journal of College and University Student Housing, 21*(1), 7–13.

Mueller, R.O., Hutchins, D.E., and Vogler, D.E. (1990). Validity and reliability of the Hutchins Behavior Inventory: A confirmatory maximum likelihood analysis. *Measurement and Evaluation in Counseling and Development, 22*(4), 203–214.

Mulaik, S.A. (1972). *The Foundations of Factor Analysis*. New York: McGraw-Hill.

Mulaik, S.A. (1987). Toward a conception of causality applicable to experimentation and causal modeling. *Child Development, 58*, 18–32.

Mulaik, S.A., James, L.R., Van Alstine, J., Bennett, N., Lind, S., and Stilwell, C.D. (1989). Evaluation of goodness-of-fit indices for structural equation models. *Psychological Bulletin, 105*(3), 430–445.

Muthén, B.O. (1988). LISCOMP: *Analysis of Linear Structural Equations with a Comprehensive Measurement Model* (2nd ed.). Mooresville, IN: Scientific Software.

Muthén, B.O. (1992). Response to Freedman's critique of path analysis: Improve credibility by better methodological training. In J.P. Shaffer (Ed.), *The Role of Models in Nonexperimental Social Science: Two Debates* (pp. 80–86). Washington, DC: American Educational Research Association.

Myers, R.H. (1986). *Classical and Modern Regression with Applications*. Boston: Duxbury.

Nesselroade, J.R., and Cattell, R.B. (Eds.). (1988). *Handbook of Multivariate Experimental Psychology* (2nd ed.). New York: Plenum Press.

Pedhazur, E.J. (1982). *Multiple Regression in Behavioral Research: Explanation and Prediction* (2nd ed.). New York: Holt, Rinehart and Winston.

Pedhazur, E.J., and Schmelkin, L. (1991). *Measurement, Design, and Analysis: An Integrated Approach*. Hillsdale, NJ: Lawrence Erlbaum.

Saris, W.E., Satorra, A., Sörbom, D. (1987). The Detection and Correction of Specification Errors in Structural Equation Models. In C.C. Clogg (Ed.), *Sociological Methodology*, 1987 (pp. 105–129). San Francisco: Jossey-Bass.

SAS Institute Inc. (1990). *SAS/STAT User's Guide: Volume 1, ANOVA-FREQ, Version 6* (4th ed.). Cary, NC: SAS Institute.

Satorra, A., and Bentler, P.M. (1986). Some robustness properties of goodness of fit statistics in covariance structure analysis. *Proceedings of the Business & Economic Statistics Section*, American Statistical Association, 549–554.

Schoenberg, R. (1987). *LINCS: A User's Guide*. Kent, WA: Aptech Systems.

Searle, S.R. (1982). *Matrix Algebra Useful for Statistics*. New York: John Wiley & Sons.

Shafer, J.P. (Ed.). (1992). *The Role of Models in Nonexperimental Social Science: Two Debates*. Washington, DC: American Educational Research Association.

Sobel, M.E. (1982). Asymptotic confidence intervals for indirect effects in structural equation models. In S. Leinhard (Ed.), *Sociological Methodology* 1982. San Francisco: Jossey-Bass.

Sobel, M.E. (1986). Some new results on indirect effects and their standard errors in covariance structure models. In N.B. Tuma (Ed.), *Sociological Methodology, 1986*. Washington, DC: American Sociological Association.

Sobel, M.E. (1987). Direct and indirect effects in linear structural equation models. *Sociological Methods and Research, 16*, 155–176.

Spearman, C. (1904). "General intelligence" objectively determined and measured. *American Journal of Psychology, 15*(2), 201–293.

Steiger, J.H. (1989). *EzPATH: Causal Modeling*. Evanston, IL: SYSTAT.

Stevens, J. (1992). *Applied Multivariate Statistics for the Social Sciences* (2nd ed.). Hillsdale, NJ: Lawrence Erlbaum.

Stone, C.A. (1985). CINDESE: Computing indirect effects and their standard errors. *Educational and Psychological Measurement, 45*, 601–606.

Tanaka, J.S. (1993). Multifaceted conceptions of fit in structural equation models. In K.A. Bollen and J.S. Long (Eds.), *Testing Structural Equation Models* (pp. 10–39). Newbury Park, CA: Sage.

Tanaka, J.S., and Huba, G.J. (1985). A fit index for covariance structure models under arbitrary GLS estimation. *British Journal of Mathematical and Statistical Psychology*, *38*, 197–201.

Tatsuoka, M.M. (1988). *Multivariate Analysis* (2nd ed.). New York: Macmillan.

Thompson, B. (Ed.). (1989). *Advances in Social Science Methodology* (Vol. 1). Greenwich, CT: JAI Press.

Thurstone, L.L. (1947). *Multiple Factor Analysis*. Chicago: Chicago University Press.

Tucker, L.R. (1955). The objective definition of simple structure in linear factor analysis. *Psychometrika*, *20*, 209–225.

Tucker, L.R., and Lewis, C. (1973). A reliability coefficient for maximum likelihood factor analysis. *Psychometrika*, *38*, 1–10.

Werts, C.E., and Linn, R.L. (1970). Path analysis: Psychological examples. *Psychological Bulletin*, *74*, 193–212.

Wiley, D.E. (1973). The identification problem for structural equation models with unmeasured variables. In A.S. Goldberger and O.D. Duncan (Eds.), *Structural Equation Models in the Social Sciences* (pp. 69–83). New York: Seminar Press.

Williams, L.J., and Holahan, P.J. (1994). Parsimony-based fit indices for multiple-indicator models: Do they work? *Structural Equation Modeling*, *1*(2), 161–189.

Wolfle, L.M. (1985). Applications of causal models in higher education. In J.C. Smart (Ed.), *Higher Education: Handbook of Theory and Research* (Vol. 1) (pp. 381–413). New York: Agathon Press.

Wright, S. (1921). Correlation and causation. *Journal of Agricultural Research*, *20*, 557–585.

Wright, S. (1934). The method of path coefficients. *Annals of Mathematical Statistics*, *5*, 161–215.

Index

Springer Texts in Statistics *(continued from page ii)*

Santner and Duffy: The Statistical Analysis of Discrete Data
Saville and Wood: Statistical Methods: The Geometric Approach
Sen and Srivastava: Regression Analysis: Theory, Methods, and Applications
Whittle: Probability via Expectation, Third Edition
Zacks: Introduction to Reliability Analysis: Probability Models and Statistical
    Methods